Origin

Origin

A Genetic History
of the Americas

Jennifer Raff

TWELVE

NEW YORK

Twelve
Hachette Book Group
1290 Avenue of the Americas, New York, NY 10104
twelvebooks.com
twitter.com/twelvebooks

First Edition: February 2022

Twelve is an imprint of Grand Central Publishing. The Twelve name and logo are trademarks of Hachette Book Group, Inc.

The publisher is not responsible for websites (or their content) that are not owned by the publisher.

The Hachette Speakers Bureau provides a wide range of authors for speaking events. To find out more, go to www.hachettespeakersbureau.com or call (866) 376-6591.

Library of Congress Cataloging-in-Publication Data

Names: Raff, Jennifer, author.
Title: Origin: a genetic history of the Americas / Jennifer Raff.
Description: First edition. | New York: Twelve, Hachette Book Group,
    February 2022. | Includes bibliographical references and index. |
Summary: "From celebrated genetic anthropologist Jennifer Raff comes the
    untold story—and fascinating mystery—of how humans migrated to the
    Americas"—Provided by publisher.
Identifiers: LCCN 2021031909 (print) | LCCN 2021031910 (ebook) |
    ISBN 9781538749715 (hardcover) | ISBN 9781538749708 (ebook)
Subjects: MESH: Genetics, Population | American Native Continental Ancestry
    Group—genetics | Whole Genome Sequencing—methods | Paleontology—
    methods | DNA, Ancient—analysis
Classification: LCC QH455 (print) | LCC QH455 (ebook) | NLM QU 450 |
    DDC 576.5/8—dc23
LC record available at https://lccn.loc.gov/2021031909
LC ebook record available at https://lccn.loc.gov/2021031910

ISBNs: 9781538749715 (hardcover), 9781538749708 (ebook)

Printed in the United States of America

LSC-C

Printing 1, 2021

For Colin

Contents

Land Acknowledgment Statement

This book was written on land taken from the Kaw (Kansa), Osage, and Shawnee nations.

Many tribes were forced into and out of Kansas prior to statehood. Today the State of Kansas is home to the Prairie Band Potowatomi Nation, the Kickapoo Tribe in Kansas, the Sac and Fox Nation of Missouri in Kansas and Nebraska, and the Iowa Tribe of Kansas and Nebraska. Because Lawrence, Kansas, is the location of Haskell Indian Nations University (formerly the United States Indian Industrial Training School, which opened in 1884), many American Indian and Alaska Native people from across the United States have ties to the region.

Summer Solstice

In our retellings I suppose we don't much bother
Keeping straight the bent details, crooked roads
In one tale after another, how we handed down
Sidelong versions of whatever happened next
Under ebbing oceans an ancient underground
Somewhere in the receding past they kept saying
Their slippery sense of community mattered, it
Shaped them, their history, the story they filled
Themselves with every day, waking their minds
Connecting to the history of memory as if it all
Felt real, seemed specific enough, logical enough
Those changing details that give rise to the world
In our retellings of the tale along a crooked road

—Roger Echo-Hawk (Pawnee historian)

Introduction

For ten thousand years, a cave on the northern tip of Prince of Wales Island in Alaska* served as a resting place for the remains of an ancient man. But on July 4, 1996, paleontologists uncovered his mandible mingled with the bones of seals, lemmings, birds, caribou, foxes, and bears (1).

The cave provided an extraordinary window into the past.[†] Paleontologist Tim Heaton and colleagues were able to tell from the remains of animals dating back as far as 41,000 years ago[‡] that this region[§] and others along coastal Southeast Alaska may

---

* Within what is now the Thorne Bay Ranger District of the Tongass National Forest.
† Officially designated 49-PET-408 when it became recognized as an historic property, the cave was called "On Your Knees Cave" by the cavers who mapped it. It has subsequently been named Shuká Káa Cave, and I will refer to it this way hereafter.
‡ Throughout this book when I mention a date it will almost always be "X years ago," which means "X calibrated years before present" but is a bit more accessible to the general reader. Note, however, that by archaeological convention the *present* is fixed at the year 1950 (otherwise the calibrated dates would become increasingly inaccurate every year). For every date given as "years ago," simply add years elapsed since 1950. For example, if you are reading this book in 2022, add 72 years to every date.
§ Though not perhaps the cave itself, which does not appear to have any animals deposited between about 17,100 and 14,500 years ago.

have served as a refuge for animals during the Last Glacial Maximum (LGM)—a period in which much of northern North America was covered by massive glaciers. As the Earth warmed and the glaciers receded, northern North America gradually was repopulated by animals from these refugia, as well as by species that had crossed the Bering Land Bridge (BLB), sometime toward the end of the last glaciation. The BLB connected the continents of Asia and North America until about 10,000 years ago.

The unexpected discovery of an ancient human presence within Shuká Káa Cave made it even more significant, particularly to the Tlingit and Haida peoples who have lived in the region for millennia. A flaked stone spearpoint had been found and reported to the island archaeologist Terry Fifield a week before, but it was assumed to be just a single isolated find. When the human mandible was found, however, Heaton immediately knew that there was much more to the site than previously expected. He stopped excavations and radioed Forest Service law enforcement to report it. The next morning, Fifield flew to the site by helicopter to assess the situation. Following the stipulations of the Native American Graves Protection and Repatriation Act (NAGPRA), Fifield brought the man's remains back to the Forest Service and called the presidents of the Klawock and Craig tribal councils the next day to notify them of the discovery.

Over the following week, aided by the NAGPRA specialist at the Central Council of the Tlingit and Haida Indian Tribes of Alaska (CCTHTA), Fifield and tribal leaders set up a consultation session hosted by the Klawock tribe and invited five Tlingit and Haida tribes to help decide what should be done next.*

---

* The Haida tribal councils of Kasaan (OVK) and Hydaburg (HCA) were invited to participate in the initial consultation in July 1996. They deferred

The initial reaction from the communities was mixed. Some were reluctant to disturb the human bones any further. But other community members wanted to learn what information the ancient man could reveal about the history of the people in the region. "As I remember those initial talks," Terry Fifield told me in an email, "council members wondered who this person might be, whether he was related to them, how he might have lived. It was that curiosity about the man that inspired the partnership at the beginning."

After much discussion and debate, community members eventually agreed that the scientists could continue their dig and study the ancient remains. They stipulated that excavations would immediately cease if the cave turned out to be a sacred burial site. They also mandated that the scientists were to share their findings with them before they were published and consult with community leaders on all steps taken during the research—and the community members would rebury their ancestor following the work.

The scientists involved agreed to all of these stipulations and updated the tribes regularly on their findings as the work progressed. Terry Fifield attended tribal council meetings and sought permission from the council whenever a journalist or filmmaker wanted to do a story on the site. Archaeologist E. James Dixon from the Denver Museum of Natural History developed a National Science Foundation–funded research project to excavate the cave, which also funded internships for tribal citizens to participate directly in the excavations. In subsequent years Sealaska Corporation, the Alaska Native Regional

---

to the predominantly Tlingit communities of Klawock and Craig, whose traditional lands are closer to the site. Thereafter, archaeologists worked with representatives from Klawock and Craig.

Corporation for the area, provided additional funding for internships to students working with the project.

This partnership between community members, archaeologists, and the Forest Service was fruitful. Over five seasons of archaeological fieldwork, seven human bones and two human teeth were recovered from inside the cave. All belonged to a single man. His bones were scattered and damaged by carnivores and were distributed across approximately 50 feet of a passage in sediments that had been churned up by water from a small spring. It was clear to archaeologists and community members alike that this was not the site of a deliberate burial; excavating his remains would not only help people learn more about the past, but it would also allow the communities to provide him with a respectful burial.*

Archaeologists were able to determine from the shape of the man's pelvis and teeth that he had been in his early 20s when he died. A chemical analysis of his teeth revealed that he had grown up on a diet of seafood. Artifacts at the site suggested that he (or someone else who had left them in the cave) engaged in long-distance trade of high-quality stone, which was used to make tools that were specially designed for hunting in the challenging Arctic environment. Radiocarbon dates from his bones revealed something astonishing: He was over 10,000 years old. These remains were from one of the oldest people in Alaska.†

The Tlingit maintain that their ancestors were a seafaring people who have lived in this region since the dawn of history. The discovery of this man, whom the Tlingit called Shuká Káa

---

* He was reburied on September 25, 2008.
† He was *the* oldest person then known (in 1996). Since then, children buried at the Xaasaa Na' (or Upward Sun River) site in the Tanana Valley in Central Alaska (Eastern Beringia) have been found that date to about 11,500 years ago. We will talk about them in chapter 6.

("Man Ahead of Us"), was consistent with oral histories that they descend from an ancient, coastally adapted people who engaged in long-distance trade. As the project progressed, the idea that this man could be their ancestor—or at least lived in ways similar to those of their ancestors—grew increasingly plausible.

Shuká Káa's story didn't end with the archaeological examination of his remains. Prior to his reburial in 2008, the tribes allowed geneticists to sample a small portion of his bones for DNA analysis. Initial tests showed that the man belonged to a maternal lineage that is very uncommon in present-day Indigenous communities, suggesting that contemporary people in the region may not be direct descendants of Shuká Káa's population.

But there's been another twist to this story over the last few years. A technological revolution has taken place within the field of paleogenomics—the study of ancestral genomes—allowing the reconstruction of an ancient person's complete nuclear genome from small samples of bone or tissue. This development allowed researchers (again with permission from the tribes) to reexamine Shuká Káa's DNA on a vastly more detailed level than the original study. His complete nuclear genome, which includes all the DNA in his chromosomes, showed that his people *were* the ancestors of present-day Northwest Coast tribes after all, again reaffirming their own oral histories (2).

Since the publication of Shuká Káa's genome, the Tlingit have continued to use genetics as a tool for studying their clan and moiety systems,* finding additional places where their lineage (as

---

* A moiety is a descent group that dictates marriage rules. In Tlingit society there are two moieties—Raven and Eagle/Wolf—which contain numerous clans. Membership in a clan is determined by matrilineal descent.

revealed by DNA), archaeological evidence, and the clan histories preserved in their oral traditions (3) speak with a unified voice.

For archaeologists, Shuká Káa added a significant piece of evidence against an outdated theory: the idea that a human presence in the Americas was recent, resulting from an overland migration about 13,000 years ago. This may have been the story you learned in school.

But we have learned over the last few decades that this story is not accurate. It does not even come close to accounting for the piles of new evidence that have been amassed by archaeologists and geneticists.

The old theory is clearly out of date, but the history of how people first got to the Americas remains a mystery, a complex puzzle to be solved. In this book, we will follow archaeologists as they draw connections between different sites across the Americas. Looking at the genetic evidence, we will examine the ways in which DNA has challenged and changed our understanding of Native American history, with a special focus on the events that are only indirectly understandable with the archaeological record. We will join scholars of both disciplines in their struggles to integrate these different clues into new models for how humans first arrived in the Americas. As we'll see later in this book, many archaeologists and geneticists now believe that people were present in the Americas far earlier than was previously thought: perhaps by 17,000–16,000 years ago, or even as early as 30,000–25,000 years ago, and that the peopling of the continents was a complex process.

At the same time as we discuss the results and models from Western* scientific approaches, it's important to acknowledge

---

* I don't especially like this term—after all, who is geographically more "Western" than the Indigenous peoples of the Americas? But I want to that Indigenous peoples of the Americas have diverse oral histories of their own origins. These traditional knowledges—like the Tlingit's understanding of their origins and their relationship to Shuká Káa—convey important lessons about the emergence of their identities as people and their ties to the land; they may or may not agree with the models presented in this book.

Histories of the Americas written by non-Native scholars tend to be dominated by the story of how Europeans colonized the continents. In the stories of Christopher Columbus reaching San Salvador, or the Pilgrims founding Plymouth Colony, or Hernán Cortés conquering the Aztecs, Native Americans are often relegated to marginal roles as supporting characters, bystanders, victims, or antagonists. Precontact histories of Indigenous peoples are given far less prominence, and many of those that do exist in popular culture are rife with outdated scholarship (at best) or blatant pseudoscience (4). With some notable exceptions, Native Americans preserved their histories in oral, rather than written, stories. European colonists did not view these oral traditions as equivalent to their own histories.

In these frameworks, Native peoples are marginalized or forgotten, excluded from public conversations, and portrayed as inhabitants of the past rather than contemporary members of society. Their own knowledge too often is ignored by non-Native scholars. This ultimately contributes to the erasure or marginalization

---

differentiate it from Indigenous sciences, which have separate origins, histories, and epistemologies. "Western science" and "Indigenous science" are not mutually exclusive, nor in opposition, but there are important differences between them. Throughout this book, my focus is on Western scientific perspectives, and the reader should understand that when I use the term *science* it is a shorthand for this framework.

of Indigenous peoples in society at large. The contributions of Native artists, politicians, writers, traditional knowledge keepers, and scholars are unappreciated. Indigenous knowledge, sacred practices, and regalia are appropriated and commodified by white people. In some cases, academics repackage and re-interpret traditional knowledge as their own scholarship without credit to Native experts.

None of this marginalization is accidental. Since the beginning of colonialism in the Americas, Native peoples have been removed, enslaved, or eliminated from their lands in order to make way for settlers. One way for colonizers to justify their claims to Native lands was to portray them as empty. The Native peoples who did remain were characterized as "savage" and backward, in need of the "civilizing" that the settler nation could provide. Disregarding or expunging Native histories from the broader narrative has been a crucial part of the larger strategy to discount the validity of age-old Native rights to lands the settlers wanted. Sadly, this practice of historical marginalization continues into the present day; as we shall see later in this book, DNA has been increasingly used as a tool for promoting narratives that disenfranchise Native peoples.

A greater awareness of the histories of Indigenous peoples on the American continents—that gives as much weight to the time *before* 1492 as after it—won't fix these issues alone. But it is an important step in itself.

This book covers a small but exciting piece of the vast and complex arc of Indigenous histories in the Americas: the very beginning, when people first came to these continents. Thanks to information we have learned both from the archaeological record and the genomes of ancient peoples like Shuká Káa, the way scientists think about this event has changed radically in recent years.

We are living through a revolution in the scientific study of human history. Geneticists and archaeologists have been working together for decades to learn from the histories archived in DNA of both present-day and ancient peoples. But because of recent technical developments in approaches for recovering and analyzing that DNA, our ability to ask and answer questions about the past has improved dramatically. New results—some surprising, others that confirm long-standing ideas about the past—are piling up at a rate so fast it's hard even for experts to keep up with each new discovery.

In the Americas the revolution has upended a long-standing model that describes the final steps that humans took on their journey from Africa across the globe. As I mentioned earlier, scientists once thought that the peopling of the Americas occurred around 13,000 years ago, following the last ice age, when a small group of people crossed the Bering Land Bridge from northeast Asia to northwestern Alaska. From Alaska they were thought to have traveled southward through a corridor that had opened up between the two massive ice sheets that blanketed northern North America. On their journey, these intrepid travelers invented new stone tool technologies for surviving in the novel environments they encountered. These technologies, which include a distinctive kind of stone spearpoint called a Clovis point, appear widely across the North American continent 13,000 years ago. The conventional model for explaining their appearance suggested that the people who made them migrated very quickly across the Americas once they passed the ice sheets.

We know today that this scenario—which dominated American archaeology for decades—is wrong. People had already been in the Americas for thousands of years by the time Clovis tools made their appearance. The updated story of how humans arrived here is still being assembled piece by piece, from clues left all over the continent: deep below the surface of a muddy pond in Florida, within the genome recovered from a tooth in Siberia, in layers of dirt baked by the hot Texas sun.

But as in the movie *Clue*, where the same events could be explained by multiple narratives, these pieces of evidence seem to tell different stories to different groups of scholars. In this book, we will examine these pieces of evidence and the various ways in which they are interpreted. We will focus our discussion primarily on the clues written in DNA, and how they support or cast doubt on interpretations of the archaeological record. A picture is gradually coming into focus, but there are still many unanswered questions.

The story of Shuká Káa and the other ancient peoples who were the first inhabitants of the Americas is not just ancient history. It's also a story about the present: Shuká Káa became the nexus of an extraordinary collaboration between different groups of people who came together to study him. This shows us how much a collaboration between Indigenous peoples, scientists, and government agencies can achieve when following an approach respectful of tribal sovereignty and values, Indigenous knowledge, *and* scientific curiosity. But in the story of American anthropology and genetics, this partnership has historically been the exception, not the rule. Fortunately, as we shall see, this is changing.

So while this book is about how scientific understandings of the origins of Native Americans have changed, we cannot tell that story without also scrutinizing how scientists have *arrived* at these understandings. This is not a pleasant history to recount.

The Indigenous inhabitants of the Americas have been treated with disrespect, condescension, and outright brutality by a number of scientists who have benefitted at the expense of the people they were so curious about. This is the legacy that contemporary anthropologists, archaeologists, and geneticists need to confront head-on; there can be no honest progress in the scientific study of the past without acknowledging those threads of human history we have dismissed, neglected, or erased in the past. The journey to knowledge has to involve self-scrutiny; scientific progress cannot be divorced from the social context in which it takes place.

These three themes—the histories reconstructed from genetics and archaeology, the story of how we achieved this knowledge, and the broader cultural questions that are raised by the research conducted in the field—are inexorably intertwined; you can't understand the whole story by examining any one of them in isolation. But just as our revolution in ancient DNA *methodologies* has allowed us to understand new histories written within the strands of DNA, it's my hope that by looking at the example of Shuká Káa, listening to scientists and Indigenous scholars and community leaders, we can transform our approaches to investigating the past.

In part 1 of this book, I'll examine the history of attempts by Europeans to understand the origins of Native Americans and explain how this fascination was born out of colonialism. Chapter 1 will discuss how Europeans grappled with the fact that the peoples they encountered in the Americas were not mentioned in the Bible. The Indigenous peoples of the Americas were an existential threat as well as an impediment to colonization; attempts to understand their origins were informed partially by curiosity and partially by a desire to subvert the threat they posed. The present-day disciplines of archaeology and biological anthropology in the United States emerged from those early attempts; ideas about racial categorization and eugenics have roots in the period as well. We will take an unflinching look at how these different roots are intertwined and their influence on subsequent research on Native American origins. We will also examine the Mound Builder hypothesis and other mythologies designed to obfuscate the truth about Native Americans as the first peoples of the Americas, as well as the other ways in which people start from the wrong place in thinking about Native American origins in the present day.

In chapter 2, I'll tell you about the "Clovis First" model of Native American origins that dominated much of 20th-century archaeology, as well as the archaeological evidence that ultimately refuted it. We'll then examine the evidence for alternative models for how people reached and dispersed through the Americas. In chapter 3, we'll look closely at the early archaeological record of Alaska. Alaska is thought to have served as a gateway through which people entered North America, but the archaeological sites there that date to the late Pleistocene and early Holocene seem to contradict the story told by sites elsewhere in the Americas. Some archaeologists believe that the evidence from Alaska supports a new version of the Clovis First hypothesis; we will examine and evaluate this model.

In part 2 of this book, we will focus on how paleogenomics—the science of learning histories from ancient genomes—has changed our understanding of the past, beginning with new findings about Mesoamerican and South American histories from sequenced genomes in chapter 4. In chapter 5 I will take the reader into our laboratory at the University of Kansas and provide a glimpse into what it's like to work with ancient DNA, explaining how we learn about population history from the fragments of ancient genomes that we coax out of samples.

In part 3, we will go through the stories that genetics has told us. I will describe what we've learned from the genomes of ancient and contemporary Indigenous peoples of the Americas and Asia, and how these stories may align with archaeological evidence for the peopling of the Americas. I'll try to make the models produced by genetics and archaeological evidence more vivid with a series of narrative vignettes that illustrate what we know about the lives of ancient people in Asia and Beringia (chapter 6), North and South America (chapter 7), and the peopling of the North American Arctic and the Caribbean (chapter 8). We will then return to the theme of how scientists obtain data in chapter 9, with a focus on how outdated models and research approaches continue to cause harm to Indigenous communities. And finally we will close on a hopeful note, as we look at the efforts of Indigenous and non-Indigenous researchers to work together with communities in developing more ethical research approaches, as the story of the study of Shuká Káa exemplifies.

I am not myself Native American; I'm the great-great-grandchild of immigrants from Poland, Ireland, and England who came to the United States in the early 20th century in search of a better life for themselves and their children. I have no idea if my ancestors were aware of the long history of settlers dispossessing the Indigenous peoples of this place of their land and culture and even committing genocide…but I am. I'm also conscious of the equally long history of people in my profession declaring themselves the experts on other peoples' origins, lives, cultures, and histories, sometimes using despicable methods to get the data they needed. It's important that I acknowledge these two facts at the beginning of the book.

I am a scientist, and this book is about the past from a scientific perspective. The stories from genetics and archaeology offered here connect the Indigenous peoples of the Americas with the broader story of human evolution, adaptation, and movements across the globe. This view of migration, however ancient, conflicts with the understanding of some (though certainly not all) tribes of their own origins. They know that they have always existed in their lands; they did not travel from somewhere else. Some Indigenous people view their origin stories as metaphorical, useful for understanding one's place in the universe and in relation to others, but still compatible with Western science. Indeed, some Native American archaeologists have demonstrated the importance of oral traditions in interpreting the archaeological record and call for careful and analytical study of these traditions and the integration of any clues they might give us to understanding the past (5). Others accept their origin stories as literal truth: They have always been on these lands; they didn't come *from* anywhere (6). I acknowledge this conflict but will not attempt to resolve it (if it is possible—or necessary—to resolve it at all). I present history in this book from the perspective of a Western scientist, but for many Indigenous peoples this is not the whole story or the only story that should be told.

This is what I believe: The aggregate understanding of ancient history is akin to a forest with many trees. Each tree corresponds with a particular compounding set of ideas about the evidence you prioritize in building your understanding of the past (7).

There are deep differences in perspectives on the peopling of the Americas, even among scientists who nominally apply the same approaches to understanding the past. For example, as we will discuss later in this book, some archaeologists are quite conservative when it comes to evaluating evidence from early sites (those that predate 13,000 years ago). They apply an impressively rigorous standard for what constitutes a legitimate archaeological site. This framing produces a very particular view of the past. I admire their rigor, but their approach differs somewhat from my own. My own metaphorical tree is rooted in the evidence produced by genetics as a starting point.

And naturally, both of these systems of knowledge can be vastly different from that of a person who prioritizes Indigenous traditional knowledge and oral histories.

My colleague Savannah Martin, a member of the Confederated Tribes of Siletz Indians who studies health disparities and stress, explained her perspective to me this way: "As an Indigenous bioanthropologist with her own creation/origin stories, I balance the interdigitations of many different ways of knowing about my peoples' histories."

Just as the forest is healthier and more beautiful for having many different kinds of trees, I believe that these different perspectives can coexist in united appreciation of the past. And as you will see, there are places where the branches—and the roots—of the trees intersect.

How I Write about Indigenous Peoples in This Book

Before Christopher Columbus opened the floodgates for mass European colonization (and the atrocities that accompanied it), there were thousands of different nations in the Americas. There still are today. Within the United States alone, there are

574 federally recognized tribes, others who are recognized by individual states, others who don't have "official" legal status but may (or may not) be seeking recognition (or reinstatement after termination) as sovereign entities, and many individuals who aren't citizens of a tribe but who are connected to communities by kinship and culture. Many more nations, tribes, and communities exist without the benefit of recognized sovereignty or autonomy throughout the rest of the Americas, each with their own unique identity, traditions, and histories.

Genetically, Native Americans are not "a people" or "a race," any more than they are a homogenous culture or speak one language. However, in talking about the peoples of the Americas, I am constrained by the limitations of the English language, and so I will frequently use the terms *Native peoples*, *Native Americans*, and *Indigenous peoples*. (Following convention, I capitalize *Indigenous* when referring to the Native peoples of the Americas, lowercase when using it as a more generalized term.) These names are themselves used by contemporary tribal members, who also refer to themselves in various places as "American Indians," "Indians," "Amerindigenous," "Natives," and "First Nations." Archaeologists often write about the "First Americans" or the "Paleoamericans"; this usage is generally an attempt to avoid the term *Indian*, which was coined by Christopher Columbus in a vain attempt to support his initial claim that he had arrived in India. Many Indigenous peoples view that term as inaccurate and offensive. (It is important to note that some are fine with it and prefer the designation over *Native American*, which they view as a colonial term.) Some of my Indigenous colleagues are uncomfortable with the terms *PaleoAmerican*, *PaleoIndian*, and *First Americans*, and, at their advice, I tend to use *First Peoples* when I am talking about people living in the Western Hemisphere prior to European contact/colonialism. I will also use these terms to refer to the portions of the genomes of contemporary Native Americans that are inherited from those peoples. I will, without apology, change the usage of a particular term used in linguistics and archaeology to refer to an Arctic group that is viewed by many of my colleagues and community partners as a slur* (8).

As we shall see, the genetic effects of European contact were profound. Today there is no "Native American genome." Contemporary Indigenous peoples are diverse, with genetic ancestries from First Peoples, but also from populations around the world. We will discuss later in this book how genetics and ancestry testing does not give insights into the question, "Who is Native American?" today.

In general, the commonly preferred way to talk about living peoples in the Americas is to be as specific as possible: e.g., "member of X" or "citizen of Y," where X or Y refers to tribe, nation, band, or group. I will do that as often as I can here.

The peopling of the Americas is not simply an esoteric bit of science and history, important to only scholars and intellectuals. It is a story of resilience, compassion, intrepidness, adventure, and loss. As the United States is engaged in a difficult

---

* Not all Arctic peoples find the word *E*kimo*—and variations of it—to be problematic. For many, this term is how they self-identify. Some, however, view it as a slur and request that it not even be spelled out. In this book I choose, like many of my colleagues, to avoid using the term at the request of the Inuit Circumpolar Council, which represents Indigenous Arctic groups from Greenland to Chukotka. Instead I will use the terms *Inuit, Arctic peoples*, and *Alaska Natives* when talking about broader groupings of Indigenous peoples who live in the region, and more specific terms (e.g., *Iñupiat*) when appropriate.

conversation about its identity as a nation, the histories of the Indigenous peoples of the Western Hemisphere—and how they have been impacted by outsiders—need to be understood and acknowledged. One place to start is by understanding just how long the First Peoples have been here.

Origin

PART I

Chapter 1

On a July afternoon, I am walking down a tree-lined street in Granville, Ohio. The traffic is sparse in this part of suburbia, which is good, because there are no sidewalks to mar the artfully manicured vista of hedges, ferns, and flowers in each yard. Signs advertising landscaping companies and home security systems are planted discreetly within the gardens and along stone pathways leading to large houses. Less discreetly positioned are the American flags and banners proclaiming their allegiance to the Ohio State University, decorating the mailbox posts.

I hear a cardinal singing and catch sight of him taking flight as I walk under his tree, his feathers red against the quiet green of the pine. In the distance, I can hear the sound of a lawnmower and, more faintly, the distinctive clink of a golf club hitting a ball. I am inhaling the smells of summer from a gentle breeze: freshly cut grass, a whiff of honeysuckle, someone's meal grilling on the coals nearby. I feel like I'm walking through a portrait of the idealized (mostly white) upper-class Midwestern American neighborhood.

On the horizon, you can just see the bluff top on the other side of Raccoon Creek Valley. The street slopes down slightly, but is still quite high above the creek. As I continue down the street, the trees begin to grow sparser, and up ahead, in front of a large grassy hill, the road splits in two. At first glance, this plot of land looks like any other park tucked away into different corners of a hundred planned communities, maintained by homeowners association dues. This is a hill where kids might play, a place where families might consider hosting picnics, an attractive place to sneak away and snatch a few quiet hours of reading and sunbathing. If you were visiting in the winter, you would probably see kids sledding down it on snowy days; the slope is the perfect angle for a great run. A few large trees grow along the side of the hill, but otherwise it's bare. There must be a terrific view of the Raccoon Creek Valley from the crest of the hill.

As I draw closer, I notice the inevitable station dispensing dog poop bags and urging people to clean up after their pets. Next to it is what I've come here to see: a historical marker. If you're like me, you find these things irresistible, especially in a town like Granville—which technically calls itself a "village"— that is filled with meticulously preserved historical buildings.

"Upon this hill," the plaque reads, "sits one of two great animal effigy mounds built by Ohio's prehistoric people." The mound measures about 250 feet long, 76 feet wide, and 4 feet high, and according to the historical sign, it is known as Alligator Mound.

If it wasn't for the sign, you might not even recognize this as an ancient mound—a sacred place to Indigenous ancestors. It's somewhat more evident if you're standing at the top of the hill, or if you look closely at the location on Google Earth, but from the street all I could see was what appeared to be natural lumps and elevations.

Alligator Mound predates every historical building in Granville. The mound was mapped and described in the 19th century by Ephraim Squier and Edwin Davis, who reported that it was "in

4

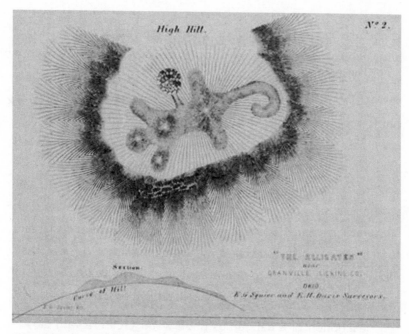

"The Alligator," Licking County, Ohio. From **Ancient Monuments of the Mississippi Valley** (1848) by Ephraim Squier and Edwin Davis, published by the Smithsonian Institution.

the shape of some animal, probably an alligator" (1), although it was very clearly *not* an alligator (and of course there are no alligators in the Midwest). Squier and Davis noted that an "altar," an elevated circular space covered in stones that showed signs of fires lit on top of it, extended via an earth causeway from the body of the creature. (When I first looked at the mound, before I read any description of it, I thought this altar and causeway was a strange extra leg on the animal.) Squier and Davis noted that Alligator Mound was one of many "works"* throughout the county, and that its location—atop the bluff—would make it extremely visible to the entire region.

---

* Short for *earthworks* or ancient constructions of earth such as burial mounds.

ALLIGATOR AND SERPENT MOUNDS

"The name historically associated with 'Alligator' mound may contain a clue to the identification of the creature represented by the effigy," wrote archaeologist Brad Lepper (2), curator of archaeology and manager of archaeology and natural history at the Ohio History Connection in a paper on the mound. He and his co-author, Tod Frolking, interpret the mound as representing Underwater Panther, one of three animal spirits—along with Thunderbird and the Horned Serpent—that are frequently depicted in the pantheon of Eastern Woodland tribes.

They note that if European settlers had asked Native Americans what the mound was supposed to depict, the description of a dangerous underwater creature with big teeth and a long tail might well have led them to believe that it was an alligator.

Underwater Panther is associated with rivers and lakes and the underworld, and begins appearing in eastern North American art around 1,040 years ago. The mound's construction has been dated to about 830 years ago.

Alligator Mound is one of two animal effigy mounds in Ohio. The other is about 80 miles southeast and also sits on a cliff overlooking a creek. Serpent Mound, in Peebles, Ohio, is an earthwork that winds sinuously for over 1,300 feet from its coiled tail to its open mouth. The serpent appears to be in the process of engulfing another oval-shaped earthwork. At the summer solstice, this head is perfectly aligned with the setting sun.

In the late 1880s, Serpent Mound was originally interpreted by Harvard University archaeologist Frederic Putnam as a serpent with an egg in its jaws. Putnam attempted to link the feature with European cultures. But work published in 2018 by Lepper and colleagues reconstructed the original dimensions of the mound and reinterpreted in non-Eurocentric terms. Their findings suggest that instead it depicts an important moment in the Dhegiha Siouan creation story: the joining of First Woman and the Great Serpent. Her acquisition of his powers through this act allowed her to create life on Earth (3).

Serpent Mound

Encountering an ancient and sacred place like Alligator Mound in the middle of a present-day upscale neighborhood is as jarring as finding a diamond ring in the debris of a street gutter. Each time I visit a mound, I am left with the disquieted feeling that comes from seeing the juxtaposition of the sacred and the mundane. When I see sled marks on the slopes of Alligator Mound in the winter, I wonder, *Are they damaging the mound by sledding? What would the ancient people who made the mound think of children playing on its slopes? What do their descendants think of this casual treatment of their ancestors' sacred place? Whose voices have been included in the interpretation of the site?* When I hear the unrelenting roars of traffic barreling down the nearby Interstates 55 and 255 as I stand atop the 100-foot-tall Monk's Mound at Cahokia near present-day St. Louis, I ponder. *What would ceremonial leaders who performed rituals here have thought of all of this?*

Serpent Mound and Alligator Mound are just two of the many ancient constructions made out of earth that once covered the region of North America archaeologists call the Eastern Woodlands: lands that lie east of the Mississippi River and south of the subarctic. These earthworks were created in a multitude of different ways and different forms. Some, like Serpent and Alligator, depicted animals or creatures. Others had high walls, stretching to enclose many acres of land in fantastically precise geometric shapes, often aligned with solstices or other astronomical markers. Some earthworks were tall and cone-shaped, found on top of bluffs overlooking river valleys or within the floodplains themselves. Some were pyramid-shaped, with flattened tops that served as platforms for ritual activities or elite dwellings. Still others were long and low, similar to the undulations of ground and grass on modern golf courses.

Mounds were often grouped together, generally reflecting

a multigenerational use of a particular location—a location chosen because it was sacred, historically significant, or simply convenient. To those of us who are trained to recognize them, mounds are visible reminders of the thousands of people who have lived, loved, warred, birthed, and died across these lands.

Earthworks that have not been destroyed or defaced by plowing, development, or looting represent just a fraction of those that originally stood throughout the Eastern Woodlands. The proximity of mounds to shopping malls, highways, houses, and parks is a fact of life in eastern North America, though many (non-Native) people are largely unaware of their presence.* And once a non-Native person *does* become aware of their presence, I hope they feel the same sense of awe at the mounds' ages and wonder as I do: Who created them? What were their purposes? What was the world around them like when they were made?

What are the histories of the people who used them?†

Many Europeans were shocked when they first realized that Native Americans were not Chinese or South Asian Indians but instead a people not described in the Bible.‡ Europeans were also curious about who had built the spectacular earthworks that were then thickly concentrated throughout the eastern reaches of the continent, testimony to a dense population.

But there was a general refusal to believe that Native

---

* All too often, the interests of economic development are deemed more important than their preservation.

† Obviously, this does not apply to their descendants, who are fully cognizant of their ancestors' histories. I ask their indulgence as I write for a broader audience here.

‡ They were also confused by the plants and animals that weren't present in Eurasia.

Americans could have made the earthworks, despite several written firsthand descriptions of Native peoples engineering and using them, as well as recorded accounts of Native Americans themselves stating that their ancestors had constructed them. Instead, Europeans fabricated elaborate mythologies to explain their presence. Most of these stories featured some version of a "lost race," fables of an "advanced" people who were wiped out by contemporary Native Americans. The bones and artifacts found within the mounds that colonizers demolished for farming were, to them, clearly the remains of this "lost race" (4).

Europeans were less unified on the exact identity of these mysterious Mound Builders. Noting the resemblance of the great platform mound at Cahokia to similar structures in Mexico, many believed that the Mound Builders were Toltecs (who were, of course, themselves Indigenous).

Alternatively, because the geometric earthworks found in Ohio vaguely resembled Early Neolithic barrows in Western Europe, they were connected to ancient peoples from that region. Or perhaps the Mound Builders were more recent: sailors led by the Welsh prince Madoc, or descendants of Irish sailors led by the monk St. Brendan.

Still others argued that the mounds were built by Phoenicians or Chinese sailors, by Romans or survivors of the lost continent of Atlantis. The Church of Jesus Christ of Latter-Day Saints, which was formed in the 19th century, believed the ancestors of Native Americans to be descendants of Lamanites, who, according to the Book of Mormon, had wiped out the god-fearing Nephites and were cursed with "a skin of blackness" as punishment. In 1901, an elder of the German Baptist Brethren Church, Edmund Landon West, suggested that Ohio—and more precisely, Serpent Mound—was actually the location of Eden as described in the Bible (5).

Mound Builder theorists of the 18th and 19th centuries emphatically agreed that the mounds were not built by the ancestors of the Native Americans they had encountered. This convenient theory allowed for settlers to believe that "Indians" were latecomers to the Americas and therefore had no legitimate right to the lands that Europeans wanted for themselves. Some pushed the idea further, with the circular logic of the colonizer, by suggesting that the "lost race" had been European.

Regardless of who was here first, it was agreed that "Indians" certainly weren't sophisticated enough to have created the extraordinary artworks that Europeans were looting from the mounds as they demolished them. By promulgating the Mound Builder myth, they disconnected Native peoples from their ancestors, accomplishments, and ties to the lands, forcing a gap into which the new settlers and their descendants happily inserted their own stories (6).

But not all Europeans and Euro-Americans accepted these narratives. José de Acosta, a Jesuit priest who lived in various places across South America and Mexico between 1572 and 1587, laid out his own theory on Native American origins in his book *Historia Natural y Moral de las Indias*. It all hinged on the assumption that Native Americans were descended from Adam and Eve. The question of whether Native Americans were human was settled—at least as far as the Catholic Church was concerned—by Pope Paul III in his 1537 encyclical *Sublimis Deus*. Catholics were informed that Indians and other "unknown" peoples not specifically mentioned in the Bible were "truly men" and should not be enslaved. It was essential instead that they should be converted to the faith by any means necessary. This did not mean that they were treated humanely by colonizers, who committed countless atrocities against Indigenous peoples, including enslaving them anyway.

Being human, therefore, these Native peoples must be descended from Adam and Eve; they must either have survived the Great Flood or (more probably, as it was written in the Bible to have covered the entire Earth), they must be the descendants of one of Noah's sons. Therefore, Acosta reasoned, they must have originally come from the "Old World," and as the chronology of the Earth was detailed in the Bible, it must not have been that long ago. He believed that they—and the remarkable animals of the Western Hemisphere—arrived by crossing some sort of land connection between Asia and North America rather than by boat across the ocean. Today we know that this land connection—the Bering Land Bridge—existed about 50,000–11,000 years ago in the center of Beringia, the lowland regions between the Verkhoyansk Range in Siberia and the Mackenzie River in Canada that remained ice-free during the last glaciation.

Of course Acosta, a man of the 16th century, never visited the Arctic regions and did not collect any field data. Instead, Acosta derived his ideas from philosophical reasoning, citations from the Bible, and the writings of Catholic saints and philosophers, rather than empirical data (7). Nevertheless, he arrived at the prevailing scientific theory of human (and nonhuman) origins on the continents centuries before the invention of contemporary archaeological or genetics methods. His ideas, far ahead of other European scholars of the time, gained very little traction for centuries thereafter. Instead, the Mound Builder myth grew in popularity.

Another narrative about the origins of the mounds came from a later and more recognizable voice. In his only published book, *Notes on the State of Virginia*, Thomas Jefferson recounted a childhood memory in which he witnessed a group of Indians visiting a mound and paying their respects to their ancestors. Jefferson devoted one chapter of *Notes* as an epic rebuttal to a popular scientific theory among European intellectuals, a theory that he saw as an existential threat to the freshly established United States. The then eminent scholar Georges-Louis Leclerc, Compte de Buffon, had asserted that the flora, fauna, and Native inhabitants of the Americas were stunted and feeble compared to their counterparts in the Old World, fueling the popularity of a kind of unified field theory of naturalism. Perhaps, he suggested, the New World botany and inhabitants had degenerated from their original Old World forms due to the prevalence of moisture and cooler temperatures throughout the continents. The deer had grown smaller, the plants more stunted, and the men weaker, more cowardly, and impotent.

By the same logic, what happened to the Indians, plants, and animals would inevitably happen to the American colonists. They would degenerate, weaken, become stunted, and their radical experiment in self-governance would never flourish. *"La nature vivante est beaucoup moins agissante, beaucoup moins forte,"* Leclerc wrote in his magnum opus, *Histoire Naturalle.* "The living nature is much less active, much less strong."

The "degeneracy theory" alarmed and infuriated many of the Founding Fathers, who viewed it as a blow against the potency of their cherished nation.* To Jefferson, it was the ultimate insult, derisive and inaccurate. His fellow statesmen pushed back in various—and highly characteristic—ways. James Madison took time from working out the foundations of the Constitution in order to catalog errors in Buffon's work, sending a long and obsessively detailed description of the American weasel to Jefferson for

---

* Most were less concerned about what it said about Native Americans.

comparison with European species. Benjamin Franklin hosted a dinner party in Paris at the home of Guillaume Thomas Raynal—one of the proponents of the theory—at which he invited both French guests and the American guests to stand and display their relative statures in order to test "on which side nature has degenerated." (The American guests were far taller than their French counterparts, although Franklin self-deprecatingly acknowledged that he was an exception.)

Jefferson himself took this fight to a completely different level, sending a stuffed bull moose to Leclerc to prove the size of America's fauna and writing a chapter of *Notes* that served as a devastating refutation to *Histoire Natural*. *Notes on the State of Virginia*, though modestly named, skillfully and passionately argued against the degeneracy theory with hard data: measurements and detailed descriptions of enormous animals and plants that far outstripped their closest European counterparts. (Jefferson cheated a bit by citing the mastodons—which he called mammoths—as an extant species, but he believed that these giant beasts were alive *somewhere* in America (8).)

Jefferson's rhetoric was particularly impassioned in the sections he wrote on Native Americans. Leclerc had characterized *"le sauvage du nouveau monde"* as impotent, unaffectionate, cold, and cowardly. Jefferson eloquently refuted each point, noting that rather "he is brave, when an enterprise depends on bravery... he is affectionate to his children, careful of them, and indulgent in the extreme... his friendships are strong and faithful to the uttermost extremity." His writing wasn't informed by extensive personal knowledge of Native Americans, but he drew upon linguistic and cultural evidence collected by others to refute Leclerc.

Jefferson's defense of the Indians was not necessarily altru-

istic, nor was it free from colonialism. Like his views on slavery,*
Jefferson's views on Native Americans were contradictory.

He was an author of the Declaration of Independence, which
referred to the Native inhabitants of the Americas as "merci-
less Indian Savages," but in his writing he asserted that he per-
sonally believed that they were equal—at least in potential—to
Europeans and should be assimilated into white society, rather
than exterminated. (He did not consider the possibility that they
should be left alone on their own lands and allowed to live their
lives according to their own traditions and laws.) Jefferson's view
of Native Americans was a common Enlightenment perspective,
perhaps best articulated by Jean-Jacques Rousseau as the concept of the "noble savage"—a romantic notion that portrayed
Indigenous peoples as primitive, close to nature, and untainted
by civilization. This perception of Indians has been built into the
mythology of the United States' origin stories. After all, the participants in the Boston Tea Party who dressed up like Mohawks

------

* I remind the reader that although Jefferson argued against the institution of slavery, he enslaved over 600 people and fathered children with
an enslaved woman, Sally Hemings, who was 14 years old when Jefferson
forced her into a sexual relationship (https://www.washingtonpost.com
/outlook/sally-hemings-wasnt-thomas-jeffersons-mistress-she-was-his
-property/2017/07/06/db5844d4-625d-11e7-8adc-fea80e32bf47_story.html).
He also believed in fundamental biological differences between the races,
and that Black people were inherently inferior to whites. Jefferson's contributions to scholarship and the discipline of archaeology in particular
are profound, and there's no denying his genius. But we must also not lose
sight of how these contributions were made. Jefferson's ability to devote
himself to a life of leisure and scholarship was a direct result of the labor
of enslaved Africans on land stolen from the Monacan people. His contributions to archaeology also contributed to the desecration of the remains
of Monacan ancestors.

had "adopted the Indian as their symbol of daring, strength, individual courage, and defiance against hopeless odds," as archaeologist David Hurst Thomas has noted (9).

Jefferson was ultimately successful in refuting the degeneracy theory with evidence from natural history, and the theory eventually dwindled into obscurity. But in the process of his debunking, Jefferson did something even more extraordinary than topple a popular scientific theory from one of the leading intellectuals of Europe. In an appendix to the text, added after an early draft, Jefferson essentially invented the American tradition of scientific archaeology by describing his excavations of a mound near his property in Virginia.

Jefferson had decided to excavate the mound in order to discover why and how it had been built. One common belief was that a mound was a burial place for warriors killed in battle. Jefferson—or, most likely, his enslaved workers—dug a trench through the center of the mound, exposing multiple layers of stone and earth containing interred skeletons and artifacts. Jefferson studied each layer of the mound in turn, recognizing what geologist Nicolaus Steno had articulated in 1669 as the *principle of superposition*: the bottom layer was the oldest, with each layer added successively on top of it. Jefferson also examined the human bones and artifacts removed from each layer. They lacked evidence of violence, Jefferson noted, and their disposition within the mound clearly indicated that the majority of them were not primary interments—bodies buried shortly after death—but rather bones that had been gathered up and reburied after the decomposition of soft tissues had taken place. He estimated the number of burials within the mound to have been close to 1,000 individuals, noting that they represented people of all ages. These facts taken together, Jefferson claimed, indicated that this mound—and by

16

inference, the many thousands of other mounds found across the Eastern United States—were not the tombs of dead warriors, but rather the common burial grounds of villages (10).

WHO BUILT "JEFFERSON'S" MOUND?

The mound that Jefferson excavated—known today as the Rivanna Mound—is one of at least 13 known to have been constructed in interior Virginia. Most of these were built on floodplains and have therefore been destroyed by erosion, as well as farming, construction, and looting. The Rivanna Mound no longer exists, and its exact location is unclear. But together archaeologist Jeffrey Hantman and members of the Monacan Indian Nation have identified the people buried within these mounds as their ancestors.

For thousands of years the Monacans, a confederation of Siouan-speaking peoples, lived throughout the piedmont and mountain regions of the state known today as Virginia. The Monacans' territories, which encompassed nearly half the state, were rich in copper, which they traded extensively with the Powhatan Confederacy along the coast and other groups to the west. They built villages on river floodplains, near the fields in which they grew corn, beans, squash, and sunflowers. They alternated seasons of farming within these villages with residence in hunting camps. They buried their beloved relatives with elaborate rituals within large mounds.

The Monacans' first experience with Europeans was probably indirect. Like other tribes across North America, they were decimated by diseases introduced by the colonists—tuberculosis, smallpox, influenza—that spread throughout the Native populations like ripples in a pond.* Even tribes in the interior that had no direct contact at all with Europeans were profoundly affected.

Shortly after settling in Jamestown in 1607, the English colonists were advised by their Powhatan trading partners that the interior tribes would be unfriendly, and the Powhatans refused to guide English expeditionary parties into the interior territories.

The few visits that the English did make to the Monacans have been poorly documented, but the overall impression that one gets from historical records and archaeological studies is that the Monacans chose to avoid contact as much as possible with the English. As one Manahoac man named Amoroleck related to John Smith, the Monacans believed that the English "were a people who came from under the world to take the world from them."

---

* However, as historian Paul Kelton (Cherokee) notes in his book *Cherokee Medicine, Colonial Germs: An Indigenous Nation's Fight Against Smallpox* (University of Oklahoma Press: 2009), warfare, slavery, and land theft perpetrated upon Indigenous peoples by colonizers created massive upheavals and devastation within Native communities, making them more vulnerable to infectious disease. Thus, he argues, the so-called "virgin soil" hypothesis is an insufficient explanation for Native American depopulation; infectious disease should be viewed as one of many interrelated factors contributing to population crashes. Along with a number of other scholars working in this area, Kelton also overturns the historical portrayal of Native Americans as passive victims of disease through his analysis of the Cherokee responses to epidemics.

The Monacans were prescient. Perhaps they had benefited from the intelligence gathered on the Spanish colonizers by a member of one of the Powhatan tribes, a man known to history as "Don Luis." Don Luis traveled with Spanish colonizers and missionaries to Mexico, Cuba, and Spain for about a decade. He used his knowledge and position to protect his people on multiple occasions; he guided a Spanish military expedition away from his homelands and only returned to his lands when on a ship with Jesuit missionaries but no soldiers. (He later assisted warriors from his tribe in killing the missionaries.) Don Luis's actions resulted in the Spanish avoiding the Virginia coast after retaliatory killings; it seems likely that he also spread word about the colonizers' practices and intentions throughout the Powhatan community and even beyond. The Powhatan, and possibly the Monacans and other allied tribes, also learned about Europeans from subsequent encounters (both peaceful and violent) with different European military, religious, and exploratory expeditions and attempts to set up colonies, such as the one at Roanoke.

However, these strategies also influenced how the English colonizers perceived the Monacans. The limited information the English had about the interior peoples led to all kinds of erroneous assumptions. For example, John Smith described the Monacans as "barbarous, living for the most part of wild beasts and fruits," and as the colonizers took over Monacan territory, they referred to it as "empty territory." But the fact that their maps designated many of these lands "Indian fields" or "Indian gardens" shows that this was fiction created to suit their purposes. Over time, the Monacans faded from Euro-American history. Jefferson's excavations may be well known to any archaeology student, but fewer have learned about the identity of the people buried in the mounds.

As colonizers took increasing amounts of land, different tribes reacted in different ways. The Monacans chose multiple strategies to deal with them but mostly continued to avoid them. As Euro-Americans encroached, the Monacans dispersed locally and isolated themselves or migrated elsewhere to join with other tribes. These strategies perhaps added to the Monacan peoples' survival into the present day; they provide a striking example of resilience in the face of tremendous adversity. Today the federally recognized Monacan Nation has over 2,300 citizens and operates programs benefiting their people and safeguarding their heritage on their reclaimed ancestral homelands on and around Bear Mountain in Amherst County, Virginia.

The Monacan tribe has worked with archaeologists and biological anthropologists to better understand the history of their ancestors buried within the mounds of Virginia.

These studies have confirmed that much of what Jefferson recorded was extremely accurate. Consistent with what Jefferson had observed, the Monacans built their mounds gradually, adding layers of earth and stone with each new burial event (11th to 15th centuries). Some of these mounds contained an enormous number of individuals: 1,000 to 2,000 estimated in Rapidan Creek (11).

Although it wasn't his primary focus, in *Notes on the State of Virginia*, Thomas Jefferson combined a detailed description of his archaeological work with ethnographic and linguistic evidence to argue that the ancient peoples who built the mounds were the same peoples living across eastern North America when Europeans first arrived. It must have taken an extremely long time for the many thousands of "Indian" languages to have developed, Jefferson argued, and they most plausibly came from northeast Asia. He even suggested a possible route for their origination:

> Again, the late discoveries of Captain Cook, coasting from Kamschatka to California, have proved that, if the two continents of Asia and America be separated at all, it is only by a narrow streight. So that from this side also, inhabitants may have passed into America: and the resemblance between the Indians of America and the eastern inhabitants of Asia, would induce us to conjecture, that the former are the descendants of the latter, or the latter of the former: excepting indeed the Eskimaux, who, from the same circumstance of resemblance and from identity of language, must be derived from the Groenlanders, and these probably from some of the northern parts of the old continent (12).

Despite Jefferson's impressive accumulation of evidence, it would be a century before this idea was accepted by the scientific community. By the 18th century, the Mound Builder hypothesis had become firmly entrenched in public opinion as the leading explanation of North American prehistory (13). Scholars and antiquarians continued to debate the identity of the

Mound Builders into the 19th century, with the majority agreeing that they were not the ancestors of Native Americans. President Andrew Jackson explicitly cited this hypothesis as partial justification for the Indian Removal Act of 1830, barely 40 years after Jefferson published his book.

> In the monuments and fortresses of an unknown people, spread over the extensive regions of the west, we behold the memorials of a once powerful race, which was exterminated, or has disappeared, to make room for the existing savage tribes (14).

Thus did the idea of Manifest Destiny become inexorably linked with concepts of racial categories. When someone asks me why I get so incensed about the concepts of "lost civilizations" and "Mound Builders" that are promoted by cable "history" shows, I simply remind them of this: In the years that followed Jackson's signing of the Indian Removal Act, over 60,000 Native Americans were expelled from their lands and forcibly relocated west of the Mississippi River. Thousands of people—including children and elders—died at the hands of the US government, which explicitly cited this mythology as one of its justifications.

As the discipline of archaeology slowly began to professionalize during the second half of the 19th century, the Mound Builder hypothesis was abandoned by most archaeologists. Subsequent archaeological studies of mounds and village sites produced an overwhelming amount of evidence that the mounds had been built by the ancestors of Native Americans. The question of who the Mound Builders were has been unanimously settled by the combined evidence of Indigenous histories, archaeology, and biological anthropology, and many mound groups are now linked with specific ancient cultures.

Jefferson's approach—direct testing by excavation and observation—previewed the best and worst of the scientific approaches in modern archaeology and physical anthropology by more than a century, and he is often referred to as the "Father of American archaeology." He brought a much-needed empirical, multidisciplinary approach to understanding the past. What we can learn from the remains of ancient peoples and the objects they left behind by following this approach has only increased over time. But Jefferson also treated the bodies of Native peoples as "specimens," viewing them as objects of study rather than as the remains of revered ancestors. This, too, as archaeologist David Hurst Thomas notes, became an ugly part of scientific traditions in the United States (15).

"To Me, He Was a Looter"

In July 1914, George Gustav Heye, a distinguished antiquarian, American Association for the Advancement of Science fellow, and life member of the American Anthropological Association, was put on trial for grave robbing.

Heye, a prolific collector of Native relics, had indeed disturbed the graves. Heye, his collaborator George H. Pepper of the American Museum of Natural History, and their crew of workmen had been excavating a mound looking for artifacts. Located on the banks of the Delaware River in Sussex County, New Jersey, near the town of Montague, the "old Minisink Graveyard" was well known to archaeologists and local people as the resting place of the Munsee Lenape, the ancestors of Delaware and Mohican Indians.

Heye's activities were not unusual for this time. "Men of science" from this era typically had free rein to remove objects and skeletons from Native American cemeteries. They used their funding to build large collections for teaching, research, and display within museums, universities, and world fairs. These collections would enable crucially important scholarship, teaching, and public education in anthropology and archaeology, which continues to this present day. Heye's own efforts (and those of the archaeologists he funded through his foundation) would form the collections of the Museum of the American Indian in 1916.* These museum collections are of tremendous value to science. But their formation caused incalculable harm to Indigenous peoples.†

Heye and his contemporaries did not consider that descendant communities' objections to the looting, plunder, and desecration of their ancestors' bodies were valid. This was largely because they prioritized the aims of scientific research, but also because they believed that it was their duty to "salvage" the bodies and

---

* About 750,000 items in the Museum of the American Indian would eventually be transferred to the National Museum of the American Indian. I have had the opportunity to visit their Cultural Resources Center, where many of these objects are housed out of view of the public, treated with reverence and sensitivity as descendant communities advise curators on their care and consult on repatriation. Although there is much more work still to do, museum curation has come a long way from the 19th century, thanks to advocacy by descendant communities and scholars, and changes in how museums themselves view their roles.

† This issue—in another context, but still quite relevant—was brilliantly illustrated by the scene in the Marvel movie *Black Panther*, in which Killmonger, the cousin and rival of T'Challa the Black Panther, challenges a white curator on how her museum "acquired" priceless artifacts from African countries during the 19th century. The scene is a pointed reminder of both the history of museum formation and a call to address present-day injustices.

objects of the "vanishing Indians" as sites and cemeteries were being destroyed by settlers' agriculture and population expansion (16).

Heye's arrest was one of the rare cases in which there were any real consequences at all for disturbing a Native American cemetery. Heye and Pepper included a discussion of the history at the beginning of their excavation report, noting that the judicial history of this case "will be of interest to future investigators of American archaeology" (17). They were charged with violating the 148th section of the New Jersey Crimes Act, which prohibited the removal of "a body of any deceased person from his grave or tomb for the purpose of dissection or for the purpose of selling the same, or from mere wantonness." Heye was convicted and fined $100* by the Sussex County Court of Special Sessions. His conviction was later overturned in 1914 by the New Jersey Supreme Court. The court noted that because the remains were not removed for the purposes of dissection, sale, or "mere wantonness," but rather for scientific study, his actions did not fall "within the purview of the 148th section of the Crimes Act." However, even while acquitting him of "mere wantonness," the New Jersey Supreme Court noted that "It may be that in what the plaintiff in error did he violated the laws of decency" (18).

Ironically, the remains of the people that Heye unearthed from the old Minisink Graveyard were actually not of any particular interest to him. He was far more interested in the funerary objects included in the grave; there were reports of exquisitely carved ornaments that had been previously unearthed from the cemetery, and Heye wanted to find more for his collection.

Because Heye wasn't interested in the skeletons that he removed from the graves, he offered them to Aleš Hrdlička, one

---

* Approximately equivalent to $2,600 today.

of the most prominent scholars of the new discipline of physical anthropology. As curator at the Smithsonian Museum of Natural History in Washington, DC, Hrdlička oversaw the creation of an enormous collection of skeletons from populations all over the world—one estimate places the total today at about 33,000 individuals (19). Many institutions across the United States formed similar collections.

These remains came from archaeologists like Heye, who donated the skeletons unearthed in their excavations; from amateur collectors who were hoping to make money; from individuals who had remains in their possession (or wished to donate their own bodies after death); from other institutions, such as medical schools that conducted dissections as part of student training; and from ethnologists who opportunistically acquired skeletons as they conducted research around the world. Other remains came from tribes and individuals who allowed Hrdlička to collect human remains from their burial sites—and they came from expeditions that Hrdlička sponsored with Smithsonian money or that he himself led. Like the museum collections of artifacts, these skeletal collections—which include the remains of people from all across the world—have formed the basis for an enormous body of research in biological anthropology and have contributed immeasurably to our understanding of past populations, human skeletal variation, human development, disease, and a myriad of other topics encompassed by the field. For example, the Terry Collection at the Smithsonian Institution, the Cobb Collection at Howard University, and the Hamann-Todd collection at the Cleveland Museum of Natural History have been used to develop methods for identifying age, stature, sex, and ancestry, useful in forensic sciences (20). They've helped researchers understand how to identify diseases and trauma that impact the skeleton, so they can better reconstruct the lives of past peoples.

But many have raised concerns about the ethics of the existence and continued use of these human remains, particularly in light of the history of their formation and who is represented in them (21).

This includes many Indigenous peoples of the Americas who did not consent to having their ancestors' remains disturbed and view their inclusion in teaching and research collections to be a violation of their traditional beliefs about the sacredness of human remains and how the dead should be treated. "They approach Indigenous remains as objects to be studied and things that have value as long as they are being used for scientific knowledge production. There is no conversation about the deep trauma and harm that can be caused by remains being exhumed, let alone being kept from repatriation, or extracting material and data out of communities without their full consent or knowledge," wrote Anishinaabe scholar Deondre Smiles in a recent essay (22).

By all accounts, Hrdlička was a mediocre archaeologist, even by early 20th-century standards. His excavation notes are careless—he didn't provide nearly enough details about context—and he discarded artifacts and destroyed ritualistically preserved ancestors to retrieve just the skulls for his collections. Like the majority of his contemporaries, Hrdlička cared very little for the wishes of descendant communities regarding their ancestors' bodies and lacked empathy or understanding of the amount of damage he was causing, either to them or the fields of archaeology and physical anthropology. While complaining about previous settlers' "visits" resulting in looting and disturbances to cemeteries, he even remarked of his own skull collecting that "[it is] strange how scientific work sanctions everything" (23).

Although there are no existing records of their reactions to Heye's and Hrdlička's activities, it's probably safe to assume that the early 20th-century Delawares and Mohicans were *not* sanguine about having their ancestors exhumed, even for scientific purposes. Certainly, present-day members of the tribes are not.

"To me, [Heye] was a looter," Sherry White, a member of the Stockbridge-Munsee Band, told me. "If somebody did that now, he'd be in jail." White was the Stockbridge-Munsee Band's tribal historic preservation officer for more than two decades and worked with members of the Delaware Tribe and the Delaware Nation of Anadarko, Oklahoma, to have their ancestors' remains and funerary objects removed from Hrdlička's collections and returned to them. They are now safely reburied in a secret location, finally at rest.*

Anthropology's Harmful Legacy of Race

Heye's acquittal gave Hrdlička free rein to study the remains from the Munsee cemetery, and he published a monograph on them the following year titled *Physical Anthropology of the Lenape or Delawares, and of the Eastern Indians in General* (24).

In this monograph, he briefly discussed the lack of evidence of disease and pathologies in this population before moving on to his true focus: measurement and comparisons.

Hrdlička sorted skulls into "type," noting that the majority of them were of one type "dolichocephaly to mesocephaly,"

---

* A number of other funerary objects were missed in the initial repatriation. The Stockbridge-Munsee tribe, Delaware Tribe, and Delaware Nation are currently working on a claim to repatriate these objects and intend to reunite them with the previous reburial.

but that a few individuals belonged to an "additional type" of brachycephalic ("broad-heads"). This classification reflected the fundamental framework through which physical anthropologists understood human variation in both the past and their present: race.

The idea that people could be classified into a few categories and ranked accordingly was deeply entrenched in early archaeology and anthropology (25). Racial categorization was a seductive method for understanding human variation because it was intuitive: According to its flawed logic, since we can easily "see" differences between people, it seemed obvious that these differences (however superficial) are reflections of some fundamental and natural truths about our species and have always been so. Scientists took the race framework as an a priori truth and sought out empirical means of proving it (therefore, ironically, creating what they assumed already existed). Even as biblical literalism gave way to an understanding of evolutionary change and deep time, racial categorization and ranking persisted in physical anthropology.

Carl Linnaeus, an 18th-century Swedish physician and botanist, made a formal description of humans as discrete biological entities in his book *Systema Naturae* (1735). In addition to developing the taxonomic classification system that biologists still use today, Linnaeus categorized people into four "types" according to a combination of physical traits, temperament, cultural practices, and behaviors: *Americanus, Europeanus, Asiaticus,* and *Africanus.** Each group had an "essential nature" that was

---

* These classifications were in place in the tenth (1758) edition of his book, which categorized them as "subspecies." In the first edition, these taxa were the less-fixed "varieties": *Europaeus albus, Americanus rubescens, Asiaticus fuscus, Africanus niger.* These concepts seem to reflect geography and skin color more than essential nature. He also described a category called

shared by all members. According to Linnaeus, people belonging to the *Americanus* type were "choleraic" (extroverted, ambitious, and energetic leaders), as well as stubborn and zealous; they "painted themselves with red lines and were regulated by customs." The hierarchical framework of Linnaeus's organization scheme reflected the concept of the Great Chain of Being, or *scala naturae*, first envisioned by Aristotle.

Linnaeus's successors grappled with residual questions stemming from his theories. Which traits were best for classifying people? Which classification schemes were most useful?

Whether people should be classified into categories was not questioned, nor was the system for ranking races by their innate qualities. To early European scientists, it was obvious that the European type was superior, and the African type was the lowest. The other types—including Native Americans—fell somewhere in between.

Measuring skull dimensions became a popular means of dividing people into racial categories fairly quickly, an approach known as craniometry.

One of its biggest proponents was the physician and naturalist Johann Blumenbach. Born in Gotha, Germany, in 1752, Blumenbach classified people as Caucasian, Mongolian, American, Malay, or Ethiopian types in his dissertation, "On the Natural History of Mankind" (1775), and sought to find a way to reconcile these different groupings with the biblical account of creation. Native Americans had long been a puzzle to God-fearing Christian European natural philosophers, because mention of their

---

Monstrosus, which covered a variety of groups of people shaped by their environment as well as mythological creatures, and a Ferus category that included wild children. See https://www.linnean.org/learning/who-was -linnaeus/linnaeus-and-race for more details.

existence was mysteriously and ominously absent from the Bible. All people on the Earth were descended from one of Noah's sons who had survived the Great Flood described in the Old Testament. Shem was the father of the Asian (Mongoloid) race, Ham the father of the African (Ethiopian) race, and Japheth of the European (Caucasoid) race.

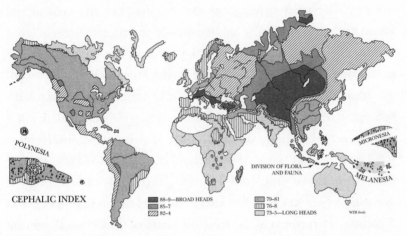

World cephalic index map (1896), adapted from *Popular Science Monthly*, volume 50

Many scholars thought that perhaps Native Americans were the descendants of Shem because of their physical resemblance to Asians. Others thought they probably weren't even human. Blumenbach was one of the first people to actually apply scientific evidence—skull measurements—toward a classification scheme and was a proponent of monogenism, the notion God created a single race—Caucasian*—and that different racial types were

---

* So named because Noah's ark must have landed in the Caucasus Mountains, and because the people in that region were the most beautiful in the world and the proportions of their crania the most balanced, according to Blumenbach.

"degenerated" from Caucasians as the result of migrating into new environments over many generations. By studying skulls, therefore, one could understand the history of humanity.

Blumenbach placed Native Americans into the Mongoloid type and suggested that they were the descendants of Asians who had migrated into the Americas in several waves.

While Blumenbach did not believe that non-Caucasians were intellectually inferior, he did believe that the Caucasians were the "most perfect" race because of the proportions of their skulls. His typological approach provided the foundation for problematic medical and anthropological research that followed.

Blumenbach's major successor—a person who Hrdlička himself called the "father of physical anthropology"—was Philadelphia physician and scholar Samuel George Morton (1799–1851). Morton believed that skulls were particularly useful for race science, since they did double duty; he thought that they both revealed not only a person's race but also their level of intelligence. It was a common assumption in the 19th century that cranial volume must be a direct reflection of intelligence: The bigger the brain, the smarter the person. (We now know that this is not true.)

Morton built upon Blumenbach's methodologies and undertook the study of crania for racial classification on a massive scale, believing that in addition to volume, a skull's shape was an essential racial marker. The cephalic index—the ratio of the maximum width to the maximum length of the skull—emerged as the simplest and most popular way of categorizing people into races. People belonged to one of three groups: long-headed people (dolichocephalic), short-headed people (brachycephalic), and those whose heads were neither short nor long (mesocephalic). These three types were referred to as Negroid, Mongoloid, and Caucasoid, respectively.

Morton reasoned that calculating the average cranial sizes of populations was the best way to assess differences in intellectual capacity between the races and developed systematic methods for measuring crania to estimate their volumes. His primary method of calculating cranial volume was to fill the crania with mustard seed (and then later with lead shot) and record the amount it took to fill each skull.

On the basis of his measurements, Morton ranked Blumenbach's types according to intelligence, with Caucasians at the top and Ethiopians at the bottom. Morton's research on the natural inferiority of non-Caucasians was explicitly used to justify slavery and the theft of land from Native Americans (26).

But Morton disagreed with Blumenbach about the origins of the races. Morton believed in polygenism, which explained human variation as the result of the separate creation of each race, rather than their eventual formation from the dispersal of the (initially Caucasian) descendants of Noah's sons. He was convinced that the differences in cranial size and shape extended deep into the past of each race; the Great Flood was simply too recent to account for all these differences. If racial traits were fixed and unchanging, the implication was that different races were actually separate species.

Another major implication of polygenism was outlined by naturalist Jean Louis Rodolphe Agassiz, a contemporary of Morton's. Agassiz postulated that different species were created in different regions of the world according to which climates were most suitable for them, and thus did not—could not—move very far from their original homelands. The same was true, Agassiz argued, of humans: Each race was created individually and separately on its own continent, and migration was the exception, not the norm, in human history.

Therefore, understanding the origins of each race could help scientists understand the history of humanity. Morton himself was particularly interested in the racial origins of Indians. In his best-known work, a study of Native American morphology called *Crania Americana* (1839), Morton suggested that Native Americans were

> marked by a brown complexion, long, black, lank hair and deficient beard. The eyes are black and deep set, the brow low, the cheek-bones high, the nose large and aquiline, the mouth large and the lips tumid [swollen] and compressed. The skull is small, wide between the parietal protuberances, prominent at the vertex, and flat on the occiput. In their mental character the Americans are averse to cultivation, and slow in acquiring knowledge; restless, revengeful, and fond of war, and wholly destitute of maritime adventure (27).

This description is remarkable to 21st-century readers for its bigotry, of course, but also for its curious combination of physical and nonphysical traits. Today our folk classifications of races tend to be based mainly on physical features: skin color, hair color, the shape of the eyes and nose. Terms like "averse to cultivation" sound very strange to a modern reader (although if you dive a bit deeper into the racist cesspool you will certainly still encounter claims about IQ and personalities and so forth).

But to a 19th-century medical man with an anthropological bent, this blending of physical and nonphysical traits was central in understanding how humans were divided into races and how those races were ranked relative to each other. It became the central preoccupation of physical anthropology in the United

States, a new discipline founded by a small group of scholars, including Hrdlička (28).

Physical anthropology's early focus was heavily influenced by the social and political issues of the time. In the first half of the 20th century, eugenics—a movement to "better" humanity by controlled breeding—was at the forefront of American culture. Some—though not all—of the early founders of physical anthropology viewed their work as crucial to the eugenics project (29).

Once evolution supplanted biblical stories of human origins, the racial framework morphed to accommodate it. Now the "savage races" were viewed as representative of earlier stages of human evolution, and useful, therefore, for reconstructing the "progress" of humanity.

But while Hrdlička and his like-minded colleagues were operating under the assumption that skull shape—and specifically the cephalic index—was a very stable, fixed marker of ancestry and useful for racial characterizations, other craniometric research was undermining this theory. Franz Boas, an anthropology professor at Columbia University, found that the cranial index actually differed between the children of Eastern European immigrants to the United States and children of similar age from their home countries (30). This demonstration of the effect of the environment on what was supposed to be a fixed trait undermined the utility of racial categories in physical anthropology, although it wouldn't be until after World War II—during which physical anthropologists and geneticists contributed to the horrors put into practice by Adolf Hitler in his "Final Solution" (31)—that a racial framework was (mostly) abandoned by the discipline (32).

The research of William Montague Cobb (1904–1990), a Black physician, anatomist, and physical anthropologist, also undermined the arguments of his contemporaries (including

Hrdlička). After receiving his MD at Howard University in 1929 (around the same time that the American Association of Physical Anthropologists was established), Cobb began studying physical anthropology with T. Wingate Todd at Western Reserve University in Ohio. Todd, whose research had shown no innate differences in brain development between the races and was vigorously opposed to the racism exhibited by segments of the physical anthropology community, trained Cobb in skeletal development and functional anatomy. Cobb would go on to build an extensive and highly regarded teaching and research skeletal collection at Howard University (currently curated by Fatimah Jackson). He published extensively on functional anatomy, but also conducted and published research that undermined racial typologies. In one of his most famous works, "Race and Runners," Cobb disproved the popular notion that African American sprinters, broad jumpers, and other athletes had an innate advantage owing to anatomical differences from runners of other races. "Genetically we know they are not constituted alike," he wrote. "There is not one single physical feature, including skin color, which all of our Negro champions have in common which would identify them as Negroes...In fact if all our Negro and white champions were lined up indiscriminately for inspection, no one except those conditioned to American attitudes would suspect that race had anything whatever to do with the athletes' ability" (33).

Despite the efforts of the antiracist factions of the discipline, physical anthropology had already provided "scientific support" for racial categorizations in the early 20th century. These categorizations are now firmly and perniciously rooted in the public mindset, contributing to a terrible legacy of discrimination and violence against Black, Indigenous, and other people of color.

Many physical anthropologists grappled mightily with the

concept of biological race using morphological data. Today, the extraordinarily detailed view of human variation that genetics offers us reveals the flimsiness of the typological approach (34). Early genetic studies in the 20th century—focused on mitochondrial and Y chromosome variation—showed that the racial categories so often used by early physical anthropologists did not correspond to actual patterns of genetic variation. Our ability to sequence whole genomes—the most amount of information about a person's genetic ancestry that it's possible to obtain—confirmed this: While populations vary genetically, this genetic variation does not follow patterns of racial categories articulated by Blumenbach, Morton, and others. If that statement surprises you, it is because these concepts are so deeply engrained within our culture.

Humans' DNA is 99.9% identical. It is that tiny difference—just 0.1%—along with what we loosely call the "environment"* that accounts for the variation in our remarkable outward appearance, or phenotype. Much of that genetic variation is distributed in patterns called clines, or gradually changing over geographic distance. It makes sense when you think about it: People who live closer to each other are more likely to marry and have children than people who live far apart from each other. Thus different alleles—versions of genes—are dispersed gradually following a pattern that geneticists call isolation by distance. Some traits—and underlying genes—show the effects of natural selection in particular environments. For example, in high-altitude environments natural selection operated to increase the regional frequency of genetic variants that help people withstand the effects of hypoxia. These patterns

---

* Meaning anything nongenetic that influences a person's phenotype: everything from factors that influenced embryonic development to one's nutrition as an adolescent to one's stress levels as an adult.

are then further complicated by our species' prolific histories of migration, which nearly always results in admixture. Contrary to the typological framework that underlay the work of early physical anthropologists, ancient DNA shows us that no population throughout human history has ever been "pure" in a genetic sense (35).

The combined effects of these evolutionary forces—natural selection, genetic drift, mutation, and gene flow—overlaid with our history of migration (or, alternatively, persistence in one area) and cultural practices have all influenced the patterns of human variation that we see today. By far the most variation—genetic and phenotypic—is present in populations that live in Africa, the continent where humans like us originated. Genetic variation generally decreases gradually in populations with greater geographic distance from Africa, reflecting our species' genetic legacies of migration, with additional adaptations to new environments. The categories of white, black, Asian (or Caucasoid, Negroid, and Mongoloid) do not accurately reflect these complexities (36).

But just because race is not a scientifically accurate way to categorize human variation does not mean that race isn't "real"—although it is a construction born of a specific cultural history, it is real to all of us and shapes our lives in profound ways (37). The horrific treatment of the Indigenous peoples of the Americas and their ancestors is just one example.

The Origins of Native Americans

Following in Morton's footsteps, Hrdlička believed that the study of human skeletons—particularly skulls—could answer the controversial question of the racial origins of Native Americans. He and his colleagues, including Earnest Hooton of Harvard University, thought their approach would complement archaeological approaches to this question. It would necessitate an extensive study of thousands of skeletons from tribes across the continents in order to understand their variation and work out their affinities.

As a result, museums and universities rushed to build large research collections of representative skeletons from all races, in order to measure and determine their racial origins (38).

And so Hrdlička acquired and examined the Munsee skeletons. He created table after table of measurements and classifications—noting the degree of pronouncement of the nasion depression (the depression at the ridge of the nose), the shape of the palate, the length of the humeri (upper arm bone), the bicondylar length of the femur. He painstakingly compared these observations to those of "Negros," "whites," and "Indians" of other tribes. He concluded that the Munsee remains were indeed the same race as other Lenape and Eastern tribes, but that the few brachycephalic individuals had likely married into the community from another tribe, possibly the Shawnee. He also noted quite definitely, without any reference to the evidence upon which he based his conclusion, that one individual in the cemetery was a white European.

Physical Anthropology of the Lenape or Delawares is a typical example of the research Hrdlička and other physical anthropologists conducted on thousands of skeletons across North, Central, and South America, trying to understand the origins of Native Americans (39).

The cranial studies from Hrdlička and other physical anthropologists as well as studies of contemporary Native American phenotypic variation (hair, skin, eye color) connected them generally with Asian populations. Hrdlička's studies of tooth morphology revealed a particular trait, a "shovel" shape of the incisors, which Native Americans share with East Asians. Later studies conducted by dental anthropologists linked Native Americans more specifically to Northeast Asians, with whom they share a high frequency of certain dental traits (40).

In a 1916 paper (41), Hrdlička summarized the state of physical anthropological investigations into Native American diversity and laid out an argument for their origins based on biological data:

1. Native Americans vary physically, but not that much; they clearly have a single origin.

2. There is no evidence that they evolved independently on the American continents as no premodern remains have been found and there were no "advanced anthropoid apes" on the continents for them to evolve from, only monkeys. All fossil and archaeological evidence from the rest of the world argues against the Americas being the origin of the human species.

3. Because "prehistoric man" had "primitive means of transportation," it was logical that "he could have come only from those parts of the Old World that lie nearest to America," that is, Northeast Asia.

4. Native Americans physically resemble Asians in all of their features, further supporting this theory.

5. Because Native Americans are not physically homogeneous, but fall into several "subtypes," there were probably multiple migrations out of Asia.

Which takes us full circle. It had become evident to physical anthropologists of the early 20th century that the scenario that José de Acosta had originally proposed on the basis of biblically inspired logic in the 16th century was, in fact, supported by multiple lines of biological evidence. Hrdlička himself believed that Native Americans arrived in North America via a land bridge that stretched from Siberia to North America. He conducted a great deal of fieldwork in Alaska trying to test this idea.

The Genetic Puzzle

Long before techniques were invented that allowed scientists to amplify and sequence DNA and "read" variation directly from genes, human genetic differences had to be inferred by looking at "classical" genetic markers such as blood groups and variants of other proteins (called polymorphisms). The frequencies of these markers were variable in different populations. This variation was easy to detect, and it gave geneticists an idea of the underlying genetic variation in different populations (42).

Data from these classical genetic markers were collected by the first generation of anthropological geneticists from many populations across the Americas, including from North Americans belonging to 53 tribes (43). By this point, although their approaches to informed consent were somewhat mixed, many of the first generation of anthropological geneticists sought consent for conducting their research from both individuals and communities as a whole. We will discuss the history of research ethics more in chapter 9.

Together, their studies revealed that Native Americans had

genetic variants that were unique to the Americas and widely shared across North, Central, and South America. These variants must have been present in a shared ancestral population. The studies also showed that Native American populations were genetically most similar to Siberian and East Asian groups.

Classical genetic markers were like the edge pieces of a complicated puzzle; they allowed for the rough outlines of the history of Native American peoples to be assembled but still gave only hints about what the picture contained. Archaeological and linguistic evidence also linked Native Americans to Northeast Asia sometime in the distant past.

Beginning in the late 1980s and early 1990s, anthropological geneticists imported tools from molecular biology that filled in a few more pieces of the puzzle. In chapter 5, I will walk you through the process of retrieving and sequencing DNA order to characterize a person's mitochondrial DNA lineage or haplogroup. Before these processes were invented, researchers were able to identify what haplogroup someone belonged to in a cruder way: by digesting extracted mitochondrial and Y chromosome DNA with enzymes that cut the molecules at different spots depending on a person's DNA sequence. The resulting fragments would appear in specific patterns when run out on an agarose gel. This method, known as restriction fragment length polymorphism analysis (RFLP), was performed on DNA sampled from Native Americans across North, Central, and South American populations. Later, when more refined techniques for amplifying and directly sequencing DNA were imported from molecular biology, large sections of the puzzle began to be filled in.

MITOCHONDRIAL AND Y LINEAGES
IN THE AMERICAS

Think of maternally inherited mitochondria and paternally inherited Y chromosomes as similar to a family tree: Individual lineages are related to each other because of descent from a common ancestor (a "grandparent"). Geneticists classify groups of closely related lineages—families—into haplogroups. Several mitochondrial and Y chromosome haplogroups arose on the American continents that are seen only in people of Native American descent. Just as members of a family may resemble one another in physical features, lineages belonging to a single haplogroup have the same set of "mutations" (DNA variants called single nucleotide polymorphisms, or SNPs) at certain spots in their sequence—these are used to classify lineages into haplogroups. And just as you are not identical to your grandmother, lineages within a haplogroup may have additional variation beyond the haplogroup-defining mutations; DNA bases change over generations.

Before European contact, all the Indigenous peoples in the Americas could trace their mitochondrial and Y chromosome lineages inherited along maternal and paternal lines, respectively, back to several founding haplogroups. (Today Native Americans are genetically quite diverse and may carry mitochondrial and Y lineages commonly found in other parts of the world as well.)

The founder haplogroups are direct descendants of ones present in Siberia, with additional variation that arose during the founding population's isolation and after population dispersal throughout the continents. The distribution of mitochondrial and Y lineages across the continents is not random; it reflects population history and has been used to identify events such as migration and gene flow or long-term continuity within a region. The so-called "Pan-American" mitochondrial haplogroups (A2, B2, C1b, C1c, C1d, C1d1, D1, D4h3a) are thought to have been present in the initial founding groups as they dispersed across North and South America. D4h3a (found primarily along the Pacific coast of the continents) and X2a (found only in North America) have been suggested to be markers of two migration routes, coastal and interior.

Mitochondrial haplogroups present in pre-contact First Peoples include the following:

South of the Arctic: A2
B2
X2a, possibly X2g
C1b, C1c, C1d, C1d1, C4c
D1, D4h3a
In circum-Arctic peoples: A2a, A2b, D2a, D4b1a2a1a

Geneticists sometimes use "A, B, C, D, X" as a shorthand for these haplogroups, reflecting a period of time when approaches for determining haplogroups could not distinguish between sub-haplogroups like A2a and A2b. All mitochondrial lineages commonly found in populations below the Arctic Circle share common ancestors between about

18,400 and 15,000 years ago. This close agreement suggests that they were all present in the initial founder population(s). Mitochondrial lineages within Siberian and Native American populations show that their ancestral populations became isolated from each other between about 25,000 and 18,400 years ago. From this genetic diversity, one can estimate the effective female population size of the founding population to be approximately 2,000. This is not the actual population size, but rather an estimate of breeding individuals (in this case, females). The actual population size would have been larger than that, but the total is hard to estimate. Arctic lineages show a much more recent expansion consistent with the Paleo-Inuit and Neo-Inuit migrations (see chapter 8).

Y chromosome founder haplogroups in Native Americans include Q-M3 (and its sub-haplogroups, including Q-CTS1780), and C3-MPB373 (potentially C-P39/Z30536). Other haplogroups found Native American populations, like R1b, were likely the result of post-European contact admixture (44).

The picture that genetic data revealed was incomplete and lacking many details, but it was enough to unequivocally answer the question posed at the beginning of this chapter and confirm the growing body of archaeological and linguistic evidence showing connections with northeast Asian populations. Many Native Americans possessed mitochondrial (A, B, C, D, X) and Y chromosome haplogroups (C and Q), clearly sharing common ancestry with haplogroups from Asia. These lineages also had additional genetic variation that arose after their separation from Asian lineages.

Together, mitochondrial and Y chromosome DNA from contemporary Native American populations gave a clear signal that they *were* the descendants of a population that had split from a larger group in northeast Asia and then had been isolated from other peoples for many thousands of years.

Ancient DNA researchers confirmed this model by finding the same lineages within ancient Native Americans. They found no evidence for ancestry from any other source in populations predating European contact. This finding effectively refuted the long-standing (though by now fringe) theories about the ancient Mound Builders (see "European Influences on Ancient North America?" sidebar). It confirmed the reconstructions based on dental and some skeletal traits linking the ancestors of Native Americans to Siberian ancestors.

EUROPEAN INFLUENCES ON ANCIENT NORTH AMERICA?

Alongside mainstream archaeological models for how people got to the Americas lie alternative ideas for their origins. These theories are bizarre, diverse, and fanciful, encompassing everything from the notion that the first peoples in the Americas were ancient astronauts to the absurd idea that Smithsonian curators are secretly hiding the skeletons of giants in their vaults. (I've been in these facilities and can assure you that there are no giant skeletons or hidden secrets.) While fringe theories about the past often pretend to be scientific, they don't follow scientific standards, and their theories are simply updated versions of the same Mound Builder myths that Europeans used to explain away the achievements and culture of Native Americans. For example, the popular author Graham Hancock claims that an ancient "lost" civilization (with psychic powers) was responsible for teaching Native Americans the technological skills they needed in order to create the earthworks in North and South America. The civilization was destroyed, according to Hancock, by meteorites during the Younger Dryas period. This is largely a restatement of the ideas of Ignatius Donnelly, who attributed the earthworks to an ancient civilization from Atlantis that was destroyed by a comet (45).

The theme of European insertion into Native American history is unfortunately echoed in an idea that continues to be championed by a very small group of researchers. Twenty thousand years ago, the story goes, the ancestors of the Clovis peoples came from across the Atlantic Ocean, leaving behind their caves and hunting grounds to seek a new land over 4,000 miles away. These ancestors, Upper Paleolithic peoples who lived in Western Europe between approximately 20,000 and 18,500 years ago, made leaf-shaped stone spearpoints using overshot flaking, which allowed them to make thin, sharp blades. These blades and other cultural attributes seen at sites in France and Spain have been called the Solutrean technocomplex by archaeologists. According to this theory, Solutreans carried their technique for manufacturing blades across the ocean to North America during their great journey at the height of the Ice Age, and 5,000 years later, Clovis projectile points were manufactured using the same approach (46).

It's easy to see why this story is popular with the general (non-Indigenous) public. It's an extraordinary narrative of human bravery and exploration—an event of equal excitement as the initial movement of anatomically modern *Homo sapiens* out of Africa, or the first ventures of people above the Arctic Circle. Out of this determined foray of Solutreans across a hostile ocean came the first peoples of the Americas. It has captured the imagination of people all over the world as a testament to humanity's ingenuity and survival in the face of overwhelming odds.

But it's completely untrue.

The Solutrean hypothesis, as it's called today, makes the case that the similarities in appearance and use of overshot flaking for both Solutrean and Clovis points are evidence of an ancestor-descendant relationship between the two cultures. From this case, proponents of the theory worked backward to find additional evidence connecting the two cultures and demonstrating the plausibility of a transatlantic journey during the height of the Ice Age. The result makes for a very compelling story, but it falls apart when you try to square it with archaeological and genetics evidence.

There's a gap of thousands of years between when Solutreans could have crossed the Atlantic and when Clovis points first start showing up in North America. People would have had to keep making their points in exactly the same way over this period, which is extremely unlikely;

we see again and again in the archaeological record that human technology is dynamic, not static and unchanging for that length of time. What's more, there haven't been any Solutrean sites found in the Americas that date to the intervening period (between about 20,000 years and 13,000 years ago) that contain Solutrean or Clovis-like points. Other sites from this period *have* been found, and their stone tools look nothing like Solutrean points (47).

There's no evidence that Solutreans used or made boats. Nor do we see any other cultural connections between Solutrean and Clovis as we would certainly expect if one was founded as the result of a migration. Archaeologists find it much more plausible that the Clovis peoples simply developed overshot flaking independently; no elaborate migration scenario is required (48).

But genetics has struck the definitive blow against the Solutrean hypothesis. If Clovis peoples were the descendants of southwestern Europeans, we can make an easy prediction: We'd expect to see at least traces of this ancestry present in ancient Native Americans. We don't.

Whole genomes sequenced from ancient Native Americans, including the only known ancient individual buried in association with Clovis artifacts, show that they are descended from an ancestral population with Siberian roots. As we'll talk about in chapters 5–8, from ancient genomes we have a very unambiguous (if complicated) picture of evolutionary history from a Siberian/East Asian ancestral population to the Americas. We see absolutely no genetic evidence for a transatlantic migration (49).

Is it possible that people could have migrated from Europe without leaving any genetic traces? Absolutely. We know in fact that this did happen—in 1000 CE, Norse mariners founded a settlement at Vinland, where they fought, interacted, and traded with a people they called the Skraelings. The Skraelings were actually precontact Inuit (we will talk more about them in chapter 8); they called the Norse the Kavdlunait. Archaeologists have established that Vinland was located at the L'Anse aux Meadows site in northern Newfoundland. Unambiguously Norse-manufactured items have been found at this site, as well eight structures, including an iron forge and boat-repair workshops. Recent dating at the site indicates that the Norse activity could have occurred sporadically for up to 200 years (50).

This Norse outpost left no genetic traces; no ancient Norse DNA has ever been recovered from remains in this region, nor is there any evidence of pre-Columbian gene flow in any contemporary inhabitants. The interactions between the Kavdlunait and the Inuit either were not sexual in nature or they resulted in no offspring, or their lineages didn't persist into the present day (and we simply haven't detected any ancient gene flow because of sampling bias). Some might argue, from analogy, that the same thing may have happened in the case of the Solutreans.

But—and this is a big *but*—we have utterly unambiguous evidence of Norse presence in North America from multiple lines of evidence. Different types of material culture in the archaeological record link L'Anse aux Meadows to the Norse in Greenland, there is a clear occupation history at the site, and there are even oral traditions among both Native Americans and Norse about this meeting. The Solutrean-Clovis connection rests upon a similarity in one kind of tool, without any other cultural connections, and a bunch of conjectures about what "could have happened." But science isn't built on "could haves" and "maybes." Models must be built based on evidence you have, not evidence you wish you had. The Solutrean hypothesis is lacking sufficient evidence to be considered a serious explanation for the origins of Clovis by the vast majority of archaeologists and—I'm going to be bold here—literally every credible geneticist who studies Native American history.

Some proponents of the Solutrean Hypothesis suggest that mitochondrial haplogroup X2a, found in some ancient and contemporary Native Americans from North America, might be a marker of European ancestry. Today, lineages of haplogroup X are found widely dispersed throughout Europe, Asia, North Africa, and North America. We can reconstruct their evolutionary relationships—much like you can reconstruct a family tree. Lineages present in the Americas (X2a and X2g) are not descended from the lineages (X2b, X2d, and X2c) found in Europe. Instead, they share a very ancient common ancestor from Eurasia (X2). X2a is of a comparable age to other indigenous American haplogroups (A, B, C, D), which would not be true if it were derived from a separate migration from Europe. Finally, the oldest lineage of X2a found in the Americas was recovered from the Ancient One (also known as Kennewick Man), an ancient individual dating to about 9,000 years ago and from the West Coast (not the East Coast as would be predicted from the Solutrean hypothesis). His entire genome has been sequenced and shows that he has no ancestry from European sources. There is no conceivable scenario under which Kennewick Man could have inherited just his mitochondrial genome from Solutreans but the rest of his genome from Beringians. Thus, without additional evidence, there is nothing to justify the assumption that X2a must have evolved in Europe (51).

No Europeans need to be invoked as the intellectual forces behind Indigenous technologies or cultural achievements. The true histories, evident in genetics, oral traditions, and archaeology, are exciting enough.

But even mitochondrial and Y chromosome sequences gave only a limited glimpse of history. It took the genomic revolution to start filling in the missing pieces, and we're still only partway there. With the ability to obtain whole genomes from ancient individuals, geneticists could confirm what was already pretty certain: There was a clear ancestor-descendant relationship between the ancient peoples of the Americas and contemporary Native Americans. And their line of ancestry stretched back thousands and thousands of years, eventually connecting during the Paleolithic, with cousin lineages stretching from present-day East Asians and Siberians. But before we can delve into that story, we must first understand what the archaeological record tells us about the earliest peoples in the Americas. We'll begin this exploration in the next chapter.

Chapter 2

Imagine living in close proximity to a wall of ice six times taller than the Willis Tower in Chicago. At its peak, nearly 2 miles thick, this wall would have been far taller and more impermeable—than "the Wall" from *Game of Thrones*.* It would extend to the east and the west, as far as you or anyone you know has ever traveled. What do you think you would have thought about it? Would you have wondered whether there was anything on the other side? Would you have assumed it represented the edge of the world?

We now know that such a wall existed. During the Wisconsin glaciation period (80,000 to 11,000 years ago), an ice sheet stretched from coast to coast, covering much of present-day Canada and the northern United States. This single ice sheet was formed from the fusion of two smaller ice sheets that each extended roughly from the coasts to the Rocky Mountains: the Laurentide ice sheet from the east and the Cordilleran ice sheet from the west. The Laurentide extended south below present-day Chicago, and the Cordilleran extended as far south as Seattle.

---

* In both the television series and the book *A Song of Ice and Fire*, the Wall is described as being almost 700 feet high.

To the north, these sheets covered Canada up through the borders of present-day British Columbia and Alberta.

We can't know what ancient peoples thought of this giant wall of ice, but it would have been a significant part of the landscape. It would have prevented anything—humans, plants, or animals—from moving between Canada and the Great Plains, in either direction. Generations lived and died in its shadow. For millennia, it was an impermeable barrier (1).

September 1908

George McJunkin's mind was probably not on archaeology while he and his friend Bill Gordon crossed the range, but as his horse picked its way through the debris deposited by floodwaters rushing through Wild Horse Arroyo, it was carrying him to one of the greatest discoveries in North American archaeological history. Over the last few weeks, McJunkin and the Anglo and Mexican ranch hands he supervised had been slowly tracking down and recovering the scattered cattle that had survived the Dry Cimarron River's flash flood.

The flood had nearly wiped out the entire town of Folsom, New Mexico, taking the lives of 17 people. As McJunkin dismounted to examine a damaged barbed wire fence, he might have been thinking about some of these lost souls, the neighbors and friends who had not escaped their homes before the raging water swept them downstream. Among the lost was Sarah Rook, the town's telephone operator, who had stayed at her post throughout the terrifying night, calling house after house to warn residents to evacuate. The townspeople, at least 40 of whose lives she had saved, found her body 12 miles downstream of the canyon, her headset still gripped in her hand (2).

As McJunkin worked on repairing the fence, his attention was caught by a pile of bones at the base of the arroyo. He recognized at once that they could not have belonged to one of the cows killed by the flood; these remains were old and dry, not freshly decayed. They were oddly shaped for cow bones, too. His curiosity piqued, McJunkin abandoned the fence and began to investigate.

McJunkin knew about animal bones, both from his extensive reading and from decades of experience working with horses and cattle and hunting bison. After a closer inspection, he determined that the remains he had found were definitely not those of cattle, but of a bison...and one much larger than any living creature he'd ever seen. The idea of discovering an unidentified type of bison was intriguing. He removed some of the bones from the site, and then began trying to rouse interest in other people to help investigate his find.

McJunkin's recognition of the importance of these remains was a reflection of his experience as a cowboy, a bison hunter, and a well-educated (albeit self-taught) naturalist. He was unable to get other people interested in the site, possibly due in no small part to the status ascribed to him as a Black man. Born into slavery in Texas, George McJunkin had been living as a free man since the end of the Civil War.

As an enslaved child, McJunkin had worked alongside his father in a blacksmith shop. By the time he was released from slavery at the age of 14 he was already an experienced horseman and fluent in Spanish. McJunkin left home to take up the life of a cowboy, going on cattle drives and trading his expertise in horse breaking for reading lessons. Once he was able to, McJunkin read every book he could find. He was well respected by fellow cowboys—of all races—and rose to the position of foreman on the Crowfoot Ranch (3).

His knowledge, experience, and fascination with science and the natural world meant McJunkin was positioned to be the right person in the right place to recognize the significance of his discovery: These partially mineralized bones were the remains of an extinct animal from the Pleistocene.

George McJunkin, possibly about 1907

Sadly, McJunkin died without ever knowing the sensation that his find would ultimately cause in the archaeology world. A few years after his death, Carl Schwachheim, a white man who worked as a blacksmith and amateur naturalist in a nearby town whom McJunkin had told about the remains, investigated the site.

After Schwachheim, along with his friend Fred Howarth, began to study McJunkin's find, they elicited interest from Jesse Figgins, the director of the Colorado Museum of Natural History in Denver, and Harold Cook, a paleontologist. Finally, a serious set of excavations were launched, and in May 1926, they uncovered a stone spear point, strongly suggesting that humans had been present alongside the extinct bison (4).

McJunkin's site uprooted the foundations of American archaeology. Before the 1920s, there was a bitter battle between scholars like Aleš Hrdlička, who believed that archaeological and skeletal evidence pointed to a very recent—within the last 5,000 years—entry of people into the Americas, and scholars like Jesse Figgins, who believed in a much earlier arrival—perhaps as early as 200,000 years ago. How early? It wasn't easy to determine.

Since radiocarbon dating methods for precisely determining the chronologies of sites were not invented until 1948 (and not applied to the Folsom site until 1951) (5), archaeologists had to infer age in a variety of indirect ways. One of them was typological: the theory that more crudely made stone tools must be older than more finely crafted ones. Another more discriminating approach was to assess whether any artifacts were directly associated (in undisturbed geological layers) with extinct, Pleistocene-aged fauna (like mammoths). Hrdlička employed another way of determining age based on cranial shape: Did any human crania from the Americas resemble those known to come from extremely ancient humans, such as Neanderthals or "Cro-Magnon" man, whose features were markedly different from those of contemporary humans?

Many claims of ancient American sites were advanced by archaeologists and antiquarians, only to be dismissed by scholars like Hrdlička and William Henry Holmes on the basis of these criteria. But the common assumption remained: All signs pointed to a very recent peopling of the Americas.

It was in this scholarly atmosphere that Figgins brought the stone point found at the Folsom site to Hrdlička's attention. The eminent physical anthropologist was politely interested but expressed concern that the point had been found out of context, as its original position in the stratigraphy was unknown. He advised Figgins to leave any future discoveries of stone points found at the site in situ so that they could be assessed by another group of scholars. When another stone point was found in 1927, and this time actually between the ribs of an extinct bison, Figgins followed Hrdlička's advice. He summoned paleontologists and archaeologists from multiple institutions to the site by telegram. One by one, they agreed that there was no doubt of the association between the point and the creature. However, even though the bison clearly belonged to an extinct species, they weren't exactly sure when that species went extinct; they needed a geologist to confirm that the stratigraphic layers containing the bison and stone tools were probably from the late Pleistocene (6).

GEOARCHAEOLOGY

Geoarchaeology is the application of geologic principles and methods to the solution of archaeological questions.

Early geoarchaeologists in the Americas during the late 19th and early 20th centuries did much of the work to establish chronologies for archaeological sites before the advent of radiocarbon dating by correlating stratigraphic layers with climatic events.

Contemporary geoarchaeologists use sophisticated approaches to address complex questions in the research of a particular site. First, they aim to understand the nature of the landscape and environment during a site's use. This environmental backdrop can tell us a lot about the ancient peoples' priorities and choices. For example, geoarchaeologists ask why people might have chosen to live in a specific place. Was it close to water? Did it offer access to (or control of) certain plants, tool stone, or animal resources? Were there geological features that gave protection from the weather (or rival groups of people)?

Another goal of geoarchaeology is to reconstruct *how* a site was formed. To understand site formation processes, geoarchaeologists "read" the profiles of sedimentary deposits and soils—which together make up the site's stratigraphic sequence. They place this stratigraphic sequence in the context of surrounding regions and other sites. This context provides information for essential understanding of who was using the site, when they were using it, and their activities. For example, by reconstructing the formation processes at a particular site, geoarchaeologists could tell us not only that it was abandoned 300 years after it was formed, they could potentially tell us *why* it was abandoned. They might be able to correlate the abandonment of a site with a prolonged drought that caused a local river to dry up. Combined with other evidence, one might reasonably infer that this change in environmental conditions forced the inhabitants to move to another location.

Geoarchaeologists are also able to assess how artifacts came to be deposited in a particular location. For example, a geoarchaeologist would be able to tell whether an interesting projectile point was left behind at the site of its manufacture, lost in a trash heap, or carried by a fast-moving river to a distant location. All three situations have major implications for the interpretation of that artifact's age and history.

Geoarchaeology is also essential for understanding where one might look for sites of a particular age. One area of particular interest right now within the archaeological community is identifying potential locations of pre-Clovis sites. Many of the earliest sites along the West Coast were likely flooded by rising sea levels after the LGM. But geoarchaeologists are currently looking for areas of isostatic rebound, a phenomenon in which layers of rock weighted down by glaciers rise in elevation after the retreat of the glaciers. They hope that such sedimentary deposits may contain clues to the very earliest peoples sitting on top of the rebounding bedrock thought to have traveled along the West Coast by boat (7).

"Pleistocene man" in North America had been found.

But not everyone believed it. Notably, in the 1928 article announcing the discovery in *Scientific American*, the editor included a note: "In the first two paragraphs of his most interesting article, Mr. Cook, the author, makes claims concerning the proof of the antiquity of man in America—claims which the editor regards as requiring a still larger volume of substantiation than the available evidence affords. With Mr. Cook's friendly concurrence, the present statement, in which the editor disclaims all responsibility for their inclusion, is published" (8).

Nevertheless, once the first evidence was recognized by archaeologists, more soon followed. The Folsom site discovered by George McJunkin would go on to yield dozens of extinct bison remains, many embedded with spearpoints. It overturned the established dogma that humans had not been present in the Americas earlier than a few thousand years ago and also presented archaeologists with an idea of how to locate early sites in the Americas based on their appearance. By searching for extinct animal remains, archaeologists began finding more and more evidence of the humans that had hunted them (9).

The Folsom site had profound consequences for American anthropology, as it left a major historical gap between the arrival of the earliest people and then-understood Late Prehistoric history (10). Gradually a timeline emerged. Excavations at the Blackwater Locality No. 1 site, near the town of Clovis, New Mexico, turned up the remains of many extinct megafauna—saber-toothed tigers, sloths, dire wolves, and mammoths—as well as projectile points that pointed to human groups' hunting and butchering activities in the region. Underneath the layers of the Folsom occupation, the archaeologists discovered an even older projectile point, made by people who lived earlier. The archaeologists named them Clovis points after the nearby town. Archaeologists searching for Clovis points soon found them within Pleistocene-age strata across North America.

As radiocarbon dating made it possible to establish absolute chronologies beginning in the 1950s, a new model to explain the peopling of the Americas began to take shape. This model took into account the apparently sudden and widespread appearance of Clovis people in the archaeological record of North America around 12,900 years ago, as well as the extinction of American megafauna—among them mammoths, woolly rhinoceroses, musk oxen—not quite 1,000 years later (11).

First, Clovis

"There are no sites in the Americas that predate the Clovis culture," declared the archaeologist in the front of the classroom. His tanned and deeply lined face, which attested to decades excavating outside in the hot sun, served as much a badge of authority as the air of absolute certainty in his words. I, a young undergraduate thrilled to be taking my first advanced archaeology course, was completely convinced. I scribbled down his declarations as fast as I could, knowing that I would need to regurgitate them in a blue book exam a few weeks later:

> dates from pre-Clovis layers at Meadowcroft (Pennsylvania) are contaminants...no Clovis age sites along the Pacific Coast...Monte Verde (western Chile) tool assemblage is co-mingled artifacts washed out from more recent sites upstream...all so-called "evidence" for pre-Clovis is poorly provenanced and unreliable...

Without once glancing at his notes, my professor recited the principal Clovis era sites that we were supposed to learn.

Murray Springs, a mammoth and bison kill site in Arizona. Aubrey, an occupation site from Texas. Anzick, the
only known Clovis-era burial in Montana...

Unlike the alleged pre-Clovis sites, he told us, *these* sites were
solid evidence of a culturally homogeneous group of people who
must have been the first Americans.

Their ancestors had come from Siberia, where they'd led
a rugged existence hunting mammoths and other massive Ice
Age beasts. They lived in small bands of extended families, who
ranged widely over the lands following their prey. When the land
bridge formed, they pursued the giant beasts across it into new
lands. Alaska was much like Siberia, and they soon ranged south
enough to encounter the ice wall, blocking any further progress. Did they think it was the edge of the world? If so, that belief
didn't last long.

Sometime near the end of the late Pleistocene, between
about 14,000 and 11,000 years ago (12), global temperatures had
risen enough for the ice wall to melt. A corridor began to form,
slowly opening from the north and south ends toward the middle. Once open, animals, plants, and people were free to travel
through an ice-free corridor. Humans followed swiftly on the
heels of the animals. Once past the southern margin of the ice
sheets, they encountered new lands and new environments completely unoccupied by other people.

These men* quickly adapted to this new landscape by inventing

---

* Archaeologists David Kilby and J. M. Adovasio have both confirmed my
impression that the archaeological literature up until the 1990s recognized women as having been around, but they didn't really get discussed
or considered as part of the theory because they *obviously* neither made
nor used stone tools. Nuanced discussions of women in the past became

a new form of hunting weapon. Thin, elegant, and deadly, the Clovis point was exactly the right innovation needed for successfully hunting American megafauna. The technology spread with its makers; their populations exploded as they spread across North America. They peopled the rest of the Americas in just a few hundred years.

A Clovis Point

The Clovis culture was swift to rise but short-lived. Clovis points disappear from the archaeological record about 200 years after they first appeared (13). The nomadic North American hunters were so skilled with the lethal Clovis points that just a millennium later all the megafauna—some 70 species—had been hunted to extinction. Lack of game demanded new ways of living, and people quickly adapted. The uniformity of the archaeological record—Clovis points as far as the eye could see—became regionally diverse, with people adjusting to local environments in unique ways. Together, the earliest hunter-gatherers who lived in the Americas from about 13,000 to about 8,000 years ago were collectively called PaleoIndians or sometimes PaleoAmericans by archaeologists.*

This model for Native American origins persisted for nearly 50 years, and it may be the one that you learned in school. Some

---

more common beginning in the 1990s. Other archaeologists I have spoken with dispute that women were omitted from study, but they were just not specifically studied.

* I dislike both of these terms, but they are the ones historically used by archaeologists to describe this period.

archaeologists called it the Blitzkrieg model, as a way to describe the amazing speed at which people charged all the way from Siberia to South America. Some called it Clovis First as a reflection of the prevailing sentiment among the archaeological community that Clovis represented the very first people in the Americas.[*] It was an elegant model, most archaeologists agreed, which neatly accounted for *nearly* all of the archaeological and environmental evidence.[†]

But evidence kept appearing that didn't quite work with the model. Even as better dating methods pinpointed the earliest appearance at Clovis to 13,200 years ago, the Clovis First model still couldn't quite satisfy everyone. From time to time, a maverick archaeologist would come forward to present a site that didn't fit the model; evidence that showed people were present in the

---

[*] They described the hypothesis that humans had caused the extinction of the giant Pleistocene beasts as the Overkill hypothesis. It's not necessarily a part of the Clovis First hypothesis, but it has become associated with the interior route because it fit with the idea of a wave of big-game hunters moving swiftly across the landscape. This hypothesis, first proposed by Paul Martin in 1973, is still hotly debated; many archaeologists and paleoecologists argue that the changing climate at the end of the Pleistocene was more of a factor in the extinction of the megafauna, and that the archaeological record shows only some megafauna were hunted by humans, not all. To this argument some add the critique that ancient Native American hunter-gatherers were far more responsible hunters and stewards of wildlife than credited in this hypothesis.

[†] Not all, though. Some archaeologists saw the derisive scrutiny of all pre-Clovis claims as echoing Holmes's and Hrdlička's approach in discounting all evidence not consistent with a <5,000-year-old peopling event. But the majority of the archaeological community had moved beyond whether Clovis was "first" or not—it probably was, but if it wasn't, one of the candidate sites would eventually hold up under scrutiny. Other archaeologists watching the popularity of the model worried that Clovis First was moving from a testable hypothesis to virtual dogma. They were prescient.

Americas before Clovis. These supposed pre-Clovis sites irritated most senior archaeologists, who, like my professor, *already knew* how the Americas were peopled. Like annoying pebbles working their way into the shoes of a runner, the archaeologists had to keep stopping to clear away these distractions before they could make progress on their research. My generation of students was inculcated with the belief that every single site proposed to pre-date Clovis had one or more fatal flaws. The attitude at the time, one of my colleagues told me, was basically "We know the answer. Don't bother us with data."

THE MEADOWCROFT SITE

Battering against the Clovis First barrier was the last thing that James Adovasio intended to do when he started his field school at the 36WH297 site in Pennsylvania. The Meadowcroft rockshelter, located on an outcropping of sandstone above the north shore of Cross Creek, offered eager young archaeology students the opportunity to learn how to excavate in the complex stratigraphy of a cave environment using what were at the time state-of-the-art protocols.

From successively deeper layers, Adovasio and his students recovered the remains of hearths, stone tools, and animal bones. They even found the astonishingly preserved remains of basket fragments—a particular specialty of Adovasio—plaited from bark strips. They slowly worked their way back in time: 500 years, then 1,000 years, then

5,000 years into the past. The rockshelter had been a comfortable place for people to camp while they hunted white-tailed deer and collected nuts, *Celtis* sp. (hackberries), and *Chenopodium* (goosefoot) from the surrounding woods.

It was a perfect field school project, but in the summer of 1974 it became something more: It revealed a significant crack in the Clovis First edifice and became a project that has continued into the present day. Adovasio and his students had excavated strata dating to the early Archaic period—older than 10,000 years; unusual for rockshelters in the area, but not unknown. Below the Archaic strata was a layer of rock—called breakdown by cavers—from an event in which the shelter's ceiling had collapsed. Below *that*, in layers that were ultimately dated to before Clovis, was clear evidence of human activities, including a bifacially flaked spearpoint.

As Adovasio described in his book *The First Americans*, after finding the point, "we immediately decamped to our favorite bar in town and polished off ten kegs of beer." He and the students knew they had found something significant in the unpretentious rockshelter. What they perhaps didn't appreciate at the time is how much of an uproar the find would ultimately cause.

The layers he and his students dug—properly termed Stratum IIa—would go on to yield the remains of plants, animals, and lithics that were ultimately dated to as old as 16,000 years ago, some 3,000 to 4,000 years earlier than the then-known earliest date for Clovis.* "Damn" was Ado-vasio's response as he stared at the report from the dating lab. He believed in the evidence.

This was a paradigm-shattering find, but only if it held up to the scrutiny of other archaeologists. And he knew the coming scrutiny would be blistering.

There was nothing ambiguous about human activities associated with Stratum IIa: this layer yielded dozens of stone tools, including the projectile point, small blades, and unifacial choppers and scrapers, made from raw materials brought to the site from distant places.

Adovasio and his students had excavated the site so meticulously that even the most vocal critics of the site could not find fault with their methods or the stratigraphy. Rather, the problem was with the dates themselves. They were too old, critics argued. The dates had been confirmed by multiple laboratories, including the outstanding lab at the Smithsonian, so the issue must be contamination, some (like my own professor) argued. Most likely the contamination was from the coal seams near the site, or perhaps from a few vitrinized (fossilized) pieces of wood found at the back of the cave. The coal contaminants yielded an infinite radiocarbon age, and so their inclusion would have made the dates from the lower two-thirds of Stratum IIa appear older than they actually were.

---

* Today we recognize the earliest Clovis sites as dating to between about 13,200 and 12,900 years ago (~11,050 14C years BP and ~10,800 14C years BP), but in the 1970s the Clovis "barrier" was earlier—about 11,500 to 12,000 years ago. For more on dates and how I talk about them in this book, see references for this chapter.

One can read the impatience of Adovasio growing in successive response publications over the years. The nearest actual coal outcrop, he pointed out, is about half a mile from the site. The vitrinized wood in the rockshelter was found over 20 feet away and two feet above Stratum IIa. Contaminants could have seeped upward into the shelter via groundwater, the critics rejoined. That wasn't geologically possible, nor were traces of coal particles (which are insoluble in water) ever detected in any sample from the site, Adovasio countered. How could this explanation possibly account for a large enough contamination event—which he estimated would have had to been over 35% of the sample's total weight—to have skewed the ages by as much as it did? And why, he asked, did critics accept the dates from the site as accurate—and thus uncontaminated—*except* those dating to pre-Clovis times?

Other critics looked at the reported remains of plants and animals—white-tailed deer, oak, and hickory—from the lowest two-thirds of Stratum IIa (the pre-Clovis layers) and argued that they couldn't possibly date to 16,000 years ago: a glacier had been *right there* at the time, just 50 miles or so north of the site. What was a temperate deciduous forest doing so close to a glacier?

Nonsense, Adovasio responded. Climatic reconstructions showing that only frozen tundra would have been present alongside this glacial margin did not take into account the effects of elevation on temperature. At 863 feet above sea level, the region containing the Meadowcroft

site could easily have had a patchy ecosystem, with deciduous trees growing even close to the glacier.

On and on the argument went. It continued for decades,
getting increasingly bitter and personal. It's rather uncomfortable to read the publications from this period (I cite a
number of them in the endnotes for this section, so you
can see for yourself). It's clear that there was a lot more at
stake than just the evidence.

I'm certainly not an archaeologist, so I cannot fairly
evaluate either the critiques or the rebuttals. It's clear
that the critics of the site wholeheartedly believe that the
dates are flawed. But looking back on this period from a
future in which we have many candidates for pre-Clovis, it
is also hard to avoid the impression that Adovasio was fully
vindicated.

Meadowcroft was arguably a victim of the Clovis First
dogma. Its discovery in the 1970s predated the discovery of other pre-Clovis sites, making it seem like an out-
lier, reinforcing doubts about it. Now that quite a few
sites are accepted as pre-Clovis, including the Cactus Hill
(Virginia), the Paisley Caves (Oregon), and the Buttermilk
Creek (Texas) sites, Meadowcroft's pre-Clovis layers seem
even more difficult to dismiss. A reexamination and additional dates with more refined methods of today would
probably yield more information and more confidence in
the results of the site, but Adovasio is adamant that he's
done enough; the evidence he gathered should speak for
itself (14).

70

I also learned in class that during the 1980s, linguistic evidence, morphological evidence, and very early genetics evidence were integrated into the traditional model, producing an extremely popular synthesis called the three-wave migration hypothesis. This model divided all Native Americans into three groups based on language groupings ("Amerind," "Na-Dene," and "Inuit-Aleut*"), each of which entered the Americas separately and sequentially. It was the so-called Amerinds who were descended from the Clovis peoples, with the other groups entering later. (See "The Rise and Fall of the Three-wave Migration Hypothesis" sidebar.)

A few years after taking this class, now a newly accepted graduate student, I found out how wrong my professor had been. It started with a conversation with my graduate advisor, Frederika Kaestle, when she casually mentioned that most geneticists didn't buy the Clovis First model at all. "Mitochondrial haplogroups coalesce much earlier than 13,000 years. Anyway, there are plenty of sites in the Americas that predate Clovis. Look at Monte Verde and Meadowcroft!" I said nothing, not wanting to betray how confused I was.

From Kaestle, I learned that by the time I was in high school in the late 1990s, geneticists had almost unanimously rejected the Clovis First hypothesis. By sequencing mitochondrial and Y chromosome lineages in contemporary Indigenous populations, geneticists had identified what they termed as "founding" lineages; those

---

* As mentioned in the introduction, I make this terminology change unapologetically to avoid reproducing what some of my colleagues and friends consider to be a slur.

which had been present in a population ancestral to all Indigenous peoples of the Western Hemisphere. (See the "Mitochondrial and Y Lineages in the Americas" sidebar in chapter 1.)

After they entered the Americas, the molecular clock was "set" for these ancestors' lineages. As mitochondrial genomes were passed from generation to generation along the female line and Y chromosomes along the male line, sometimes one of their DNA bases would spontaneously change: from an A to a G or from a C to a T. The changed DNA base (sometimes called a mutation or variant) could get passed down as well. Over time, as populations expanded and people moved across continents, these lineages added mutations and diversified, expanding outward like tree branches. During the 1990s, geneticists spent a lot of time examining the branching patterns of different mitochondrial "trees" (which they called haplogroups) and working backward using the molecular clock to determine the date at which all these lineages shared a common ancestor—the trunk of the tree. In genetics-speak, we call this point in time when two lineages join up at their common ancestor the coalescent event, and it is a very powerful tool for understanding many aspects of genetic histories in all organisms.

Of course, it isn't all quite so simple. Many assumptions go into correlating the coalescent events with actual dates. Geneticists must operate under the assumption that the molecular clock "ticks"—or accumulates mutations randomly—at a constant rate. Many papers have been written about whether that is an accurate assumption, and what that rate actually is. Depending on the accepted rate, the coalescence of all mitochondrial haplogroups found in the First Peoples of the Americas was estimated at either between 20,000 to 15,000 years ago or between 30,000 to 20,000 years ago (see the "Mitochondrial and Y Lineages in the Americas" sidebar in chapter 1 for a more recent and precise estimate). Neither of these was compatible with the Clovis First model of initial peopling around 13,000-ish years ago.

As geneticists were working on calculating these dates—and arguing whether the Americas were peopled in a single migration or several—archaeologists were turning up more and more convincing evidence of humans' presence in the Americas before Clovis.

THE RISE AND FALL OF THE THREE-WAVE MIGRATION HYPOTHESIS

Evidence about the past from different fields can be frustratingly hard to integrate and interpret. In 1986, when anthropological genetics was still a very young field, a group of three researchers—Joseph Greenberg (linguist), Christy Turner II (anthropological linguist), and Stephen Zegura (anthropological geneticist) attempted to reconcile the then available linguistic, genetic, and dental evidence to produce a unified model for the peopling of the Americas. Their model rested on the assumption that the Americas were peopled by anatomically modern humans from Asia after the terminal Pleistocene.

The savvy reader will notice that we have not talked about linguistics much in this book. That is certainly not because I believe that linguistic evidence can't tell us much about human history—quite the contrary! The extraordinary diversity of language families in the Americas suggests that humans have been present in the Western Hemisphere for a very long period of time.

It is generally accepted by many linguists that there is a temporal limit to the histories they can reconstruct. "Linguistics tends not to speak to the great time depth that the peopling of the Americas relates to," linguistic anthropologist Mark Sicoli told me.

This is because of the methods that most linguists typically use for reconstructing histories. Traditionally, they rely upon comparisons between the sounds of words that mean similar things in different languages to see if they have a common origin (referred to as homologous or cognate words*). This is analogous to genetics; the assumption is that the degree of lexical similarity between two languages reflects how closely related they are.

But just as with genetics, there's a great deal of complexity that can obscure historical relationships. For example, people often borrow words from other languages. Some words, called false cognates, may appear to be cognates, but actually have entirely different origins.† After a certain point, things get so messy that any inferences about historical relationships are unreliable. "The data run out between about 7,000 to 9,000 years [ago]; language

---

* Not all cognate words have the same meaning, however, as with the German *tier* meaning the class of animal and the English *deer* meaning a specific animal.

† For example: English *much* and Spanish *mucho*. While these words look alike with similar meaning, they come from different words historically.

has changed so much that people can't really go beyond the early Holocene using methods focused on the retention of shared ancestral vocabulary," Sicoli told me.

The linguistic diversity of Indigenous peoples in the Americas is mind-boggling: There are estimates that over 1,000 languages were spoken in the Western Hemisphere at the time of European contact (and that number is probably an underestimate).

In attempting to classify all the languages in relation to each other, Greenberg grouped them into three major families: Inuit-Aleut, Na-Dene, and a third group that they called Amerind. In their paper, the authors asserted that these linguistic groups corresponded to biological groupings made on the basis of dental traits and genetics. They proposed that the Americas were peopled in three waves of migration: the Amerinds first, the Na-Dene second, and the Inuit-Aleut third.

The paper was immediately excoriated by other linguists. The root of Greenberg's hypothesis, and the reason why it was controversial, was not because he "discovered" a later migration of Arctic peoples using linguistic data; that was already well understood from Arctic peoples' own oral histories, archaeological data, biological data, and linguistic data preceding Greenberg's work. Rather, Greenberg's hypothesis was controversial because he grouped all other languages (besides those spoken by Athabaskans and Arctic peoples) into the Amerind category. In a famous critique, Johanna Nicols countered that the diversity among the languages grouped as "Amerind" would require something on the order of 35,000 years to develop, not the 12,000 years Greenberg assumed to fit the Clovis First hypothesis.

The methods which he used to construct these groupings were considered highly problematic by his colleagues. Greenberg did not distinguish between homology and other factors that might cause words to resemble each other. This resulted, critics argued, in fundamental errors that were fatal to his classification scheme. Furthermore, the alleged "correspondences" between linguistic, genetic, and morphological groupings broke down when it came to specifics: The so-called Greater Northwest Coast Group identified by dental traits did not correspond to the Na-Dene linguistic grouping—it included Inuit-Aleut speakers and other people who would have been classified as Amerind by Greenberg. The genetics data did not fit either.

Despite these flaws, some geneticists eagerly adopted the model. The three-wave migration hypothesis became the standard model that all genetics studies of peoples in the Americas tested with new evidence for decades. Between 1987 and 2004, 80 out of the 100 papers published on genetic variation in Native American populations were influenced by (or mentioned) this model.

Eventually, however, mitochondrial and Y chromosome DNA produced clear patterns that did not correspond to these linguistic groupings. Despite a brief—and somewhat mystifying—revival of the three-wave migration hypothesis by a team of geneticists in 2012, the hypothesis has been utterly falsified by the genomics data.

This is a very nice example of how multiple fields can work together to test one another's hypotheses. As geneticist Emőke Szathmáry wrote in response to the Greenberg et al. paper: "May there always be creative individuals who propose models, and may there always be scientists whose testing will finally allow us to select the scenario that is most likely" (15).

Foremost among the pre-Clovis candidates was the site of Monte Verde in Chile. In the 1970s, archaeologist Tom Dillehay began to excavate at the site after the bones from gomphotheres—extinct elephant-like creatures—began turning up with what appeared to be cut marks on them. What he and his excavation team found was beyond anything they could have ever imagined. Because it had been buried under a peat bog, Monte Verde had an almost miraculously preserved set of organic and non-organic objects: ropes with knots still tied in them, the remains of wooden-framed huts, remnants of meals (including wild potatoes!), mammoth remains with the soft tissue still preserved, the remains of medicinal plants, and—perhaps most poignantly of all—the impression of a young person's footprints left in a layer of mud and clay. At 14,600 years ago, Monte Verde's archaeology predated the earliest date for the appearance of Clovis in North America by over a thousand years (16).

Monte Verde was dismissed by many archaeologists for its incompatibility with the Clovis First model's dates, for the fact that the stone artifacts that were found there looked nothing like Clovis points, and mostly for its overall strangeness. But Dillehay persisted, and in 1997 a group of eminent archaeologists traveled to the site to evaluate whether it was indeed a legitimate archaeological site and to determine if the radiocarbon dates of its strata were accurate. After visiting the University of Kentucky to view the materials that Dillehay excavated, the entire group traveled to the site to evaluate its stratigraphy.

It was, by all reports, a rather tense and contentious visit. But at the end, archaeologist David Meltzer put the question to a vote: Was Monte Verde a legitimate pre-Clovis occupation site?

The experts said yes (17).

Once Monte Verde broke the Clovis barrier, archaeologists began to rethink how the peopling of the Americas might have occurred. They took another look at known pre-Clovis sites and began to recognize new ones, including Paisley Caves (~14,000 years ago) in Oregon, Page-Ladson (~14,500 years ago) in Florida, the Manis mastodon kill site in Washington (~14,000 years ago), Huaca Prieta in Peru (~14,500–13,500 years ago), the Buttermilk Creek complex sites in Texas (~15,000 years ago), the Schaefer and Hebior sites in Wisconsin (~14,500 years ago), the Cactus Hill site in Virginia (~16,900–15,000 years ago), the Cooper's Ferry site in Idaho (~16,000 years ago), the Taima-Taima site in Venezuela (~14,000 years ago), and many others (18). Each site has its critics, and some are severe. However, despite my archaeology professor's insistence to the contrary, one can say broadly that the totality of the archaeological evidence (we will examine recent evidence from genetics in detail in chapters 5–8) indicates that humans were in the Americas by (at the most conservative estimate) 15,000–14,000 years ago, more likely between 17,000 and 16,000 and perhaps even as early as 30,000–20,000 years ago (if you accept the evidence from some of the sites in Central and South America) (19).

THE ORIGIN OF CLOVIS?

Clovis points are seen nowhere before about 13,000 years ago, but then they appear nearly simultaneously in the archaeological record across large swaths of North America. Where exactly does the Clovis technocomplex come from? It's difficult to say for certain how the Clovis point evolved, because we have conflicting evidence.

The region with the greatest number of sites—and the most diversity in fluted point form—is the southeastern United States. This has led some archaeologists to suggest that Clovis technologies were developed either there or somewhere near there—note that the Page-Ladson site in Florida predates Clovis and has evidence of human activity during the Clovis period. Unfortunately, a good chronology can't be established in the Southeast as there aren't any reliably dated Clovis points anywhere in the region; the points were all found on the surface out of context.

Another region with a high density of Clovis sites is the southern Plains. But as you move northward, Clovis sites tend to be younger in age.

One site that has been claimed to show key evidence of the origin of Clovis is the Debra L. Friedkin site in Texas, which contains a series of occupations dating from the late Archaic through pre-Clovis levels dated between 15,500 and 13,500 years ago. These pre-Clovis levels expose a fascinating sequence: stemmed points in the lowest levels, then directly above them in levels dating to about 14,000 years ago somewhat crudely shaped projectile points that were made using similar methods to Clovis points but don't have their distinctive fluting. Directly above those were Clovis points. And in each level, we can see that people were also making other kinds of tools in very similar ways: scrapers for removing hair from hides, blades for various cutting tasks, and choppers. Archaeologist Michael Waters thinks that these sites contain the clues to the evolution of Clovis: the pre-Clovis tools (which have been classified as the Buttermilk Creek complex) are the technological ancestors of the Clovis tools. "I see broad connections between Clovis and pre-Clovis sites," he told me. "The early sites show that biface, blade, and osseous [bone tool] technologies were present before Clovis across the Americas. It is from these technologies that Clovis could emerge."

Waters's interpretation of the Buttermilk Creek complex sites is disputed by some archaeologists, who question the stratigraphic integrity of the site or the dates he obtained. Others dismiss it (and other pre-Clovis sites) on the grounds that there's too much technological variation between them to make sense. *Look at how uniform the technologies of Clovis (and Western Stemmed) traditions are*, they argue. *This doesn't make sense coming out of the variation in all pre-Clovis sites.*

But should we expect to find such uniformity among early (perhaps the earliest) sites? Waters doesn't think so. "I would expect them to be different to some degree," he told me. "I would not expect a unified phenomenon like Clovis for the First People. Especially since people would be adapting and perhaps inventing new things as they go to deal with the environments they find. Clovis is something very different."

Waters admits that his work on the origins of Clovis has a long way to go. "Now to be fair, there are not a lot of pre-Clovis artifacts except at Gault and Friedkin. So we are really only getting one look at this potential connection." But he is confident that the question of Clovis's origins will be solved in time. "As we get more sites, we will know more. Look how long it took to define Clovis after it was found...20 years" (20).

Routed

The new dates for pre-Clovis sites posed new problems. We know that the ice-free corridor between the Laurentide and Cordilleran ice sheets was open at least 13,000 years ago—there's evidence for gene flow between bison populations north and south of the ice sheets occurring at this time, osteological evidence for elk migration through the corridor by about 12,800 years ago, and direct evidence of humans in the center of the corridor by 12,350 years ago (21). But was the corridor open long enough to allow people to walk through it in time to populate the pre-Clovis sites? This is a controversial question among archaeologists. Some believe that the corridor was open early—between 15,000 and 14,000 years ago (22). These archaeologists tend to be more cautious about the validity of pre-Clovis (some in this group do not accept that *any* pre-Clovis sites are valid). Other archaeologists, particularly those who accept pre-Clovis sites of 16,000 years ago or earlier, believe that the corridor couldn't possibly have been open in time for an initial peopling via that route. This perspective is bolstered by several independent lines of genetic evidence, discussed below.

Importantly, there has never been any archaeological evidence whatsoever showing that anyone moved from Beringia through the interior corridor to the Plains or Great Lakes area. We don't see the kinds of tools—microblades and a kind of spear point called a Chindadn point—that people were making at the northernmost end of the corridor at sites within the corridor or below the ice wall.* The only archaeological evidence in the corridor is of people moving *northward*: from the Northern Plains to Alaska/Yukon several millennia after Clovis (23).

A research team extracted and analyzed microfossils and pollen from sediment cores taken from two lakes—Charlie Lake and Spring Lake—which are the remnants of glacial Lake Peace, a massive lake that formed in the middle of the corridor as the ice sheets melted. By sequencing all the DNA from each layer in the sediment core, the researchers generated an overall picture of what kinds of animals and plants (and microorganisms) were living at different periods in that region's past. This molecular time capsule showed them that even if the entirety of the ice-free corridor was open by 13,000 years ago, it wouldn't have had vegetation until about 12,600 years ago and animals living within it until about 12,500 years ago. The corridor's "viability" date would have constrained the movement of people through it, as they would have needed things to eat during their trek through the 2,000-kilometer corridor (24). In addition, as paleoecologist Scott Elias notes, the melting of the enormous ice sheets would have littered the corridor with huge amounts of rock, mud debris, chunks of ice, and water everywhere as the billions of gallons

---

* We'll talk more about these in the next chapter.

of frozen water were released. "As a human migration route, it would have been absolutely awful," he told me.

The genomes of Native Americans also argue against the ice-free corridor route. We will talk about this more in later chapters, but complete genomes from ancient and contemporary Indigenous peoples show that major population splitting events were almost certainly associated with the initial peopling of the continents. These population splits occurred extremely rapidly—so rapidly that they have been described as "leap-frogging" southward across large tracts of the American landscape. This is not a pattern consistent with slower, overland diffusion of hunter-gatherer populations. Instead, it matches what one would expect to see if people were traveling by a much faster method: by boat.

THE WESTERN STEMMED TRADITION

Clovis sites, marked by fluted projectile points and other components of the Clovis toolkit, are found throughout North America.

A different toolkit is more common at early sites in the intermountain West (the region between the Cascade/Sierra Nevada and Rocky Mountain ranges). In this region, sites contain so-called Western Stemmed points that are quite distinctive from the lanceolate bifacial projectile points in their bases, where they were hafted to a spear shaft. They also made crescent-shaped knives out of stone, which are rarely seen at any Clovis site. They're made from obsidian or other volcanic rock, unlike Clovis points, which are more often made of high-quality chert. (In the Great Basin, however, many Clovis points are made of obsidian.)

Similarly, Clovis points at sites in the Great Basin, Colorado Plateau, and Columbia/Snake River Basin are either surface finds or from buried contexts that haven't yet been dated. It's difficult to derive a relative chronology for the two kinds of sites, although Western Stemmed points that are associated with radiocarbon dates at sites like Paisley Caves in Oregon, Cooper's Ferry in Idaho, and Bonneville Estates Rockshelter in Nevada are at least as old—if not older—than the earliest known Clovis sites.

The Western Stemmed sites are so different from Clovis sites that many archaeologists doubt that they were made by people with shared culture and identity. Some have argued that the Western Stemmed sites are traces of the first people to travel south of the ice sheets along the West Coast, and that Clovis sites were occupied by a slightly later, genetically distinctive population. However, that hypothesis was undermined by the recent finding that an individual from the Spirit Cave site in Nevada, associated with the Western Stemmed Tradition, was genetically similar to the individual from the Anzick site in Montana, associated with the Clovis Tradition. Although these are admittedly just two individuals, the affinities between their two genomes are not consistent with a model of separate origins for their populations. This is an excellent example of the incongruence sometimes found between cultural traditions and genetics, and a reminder that every assumption made about population relationships based on archaeology needs to be tested with biological data (25).

Moving down the West Coast by boat would have allowed people to travel faster and begin their migration earlier than through the ice-free corridor, a hypothesis first proposed by the Canadian archaeologist Knut Fladmark in 1979 (26). The Cordilleran ice sheet melted back from the Pacific coast around 17,000 years ago, meaning that people could have lived along the coast, eating kelp, waterfowl, fish, shellfish, and marine mammals, and periodically going inland for hunting and gathering animals and plants that

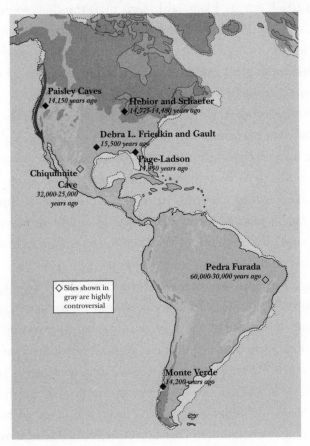

Adapted by *Scientific American* map by cartographer Daniel P. Huffman.

had survived the LGM south of the ice sheet. The Coastal Migration theory was hotly debated because the interior route hypothesis was accepted by a large number of archaeologists.

But after Monte Verde demonstrated that people must have progressed to South America by 14,600 years ago, the coastal migration theory suddenly became the best explanation for how people could have traveled south of the ice sheets in time (27).

But did the First Peoples in the Americas make and use watercraft? It seems likely, although direct archaeological evidence is scarce. We know that humans developed and employed seafaring technologies quite early; it appears from genetic and some archaeological evidence that humans were using boats to travel to Australia by 75,000 to 62,000 years ago (28).

We have no direct evidence that the *earliest* peoples in the Americas had maritime adaptations. We know that people were making and using watercraft by 13,000 years ago because of the presence of a person's remains dating from that period on Santa Rosa Island in the Channel Islands, off the coast of Southern California. It would have taken a boat to reach the site during that period, and so it seems reasonable to tentatively infer that boats were in use by his ancestors as well (29).

One important component of the Coastal Migration model was that coastal marine food sources would have been far more abundant than those of the interior route. While traveling southward through the interior of the continent would have required serially encountering and adapting to new ecosystems (mountains, deserts, plains), people traveling southward via the coast would have had reliable access to food resources with which they were already familiar. Coastal resources are fairly consistent regardless of latitude, and people would have encountered similar ecosystems along the Pacific coast from Southeast Alaska to

Tierra del Fuego. In recognition of the significant role that nutritionally valuable and abundant seaweeds might have played in a coastal dispersal, this ecological model has come to be known as the Kelp Highway hypothesis (30). While much more work needs to be done to fully test the ecological hypothesis, it's supported by the fact that people who lived at Monte Verde ate algae and seaweed over 14,000 years ago and by the dietary evidence associated with Shuká <u>K</u>áa that we discussed at the very beginning of this book.

Out of Japan?

The Kelp Highway or Coastal Migration hypothesis is widely accepted by archaeologists and geneticists for the various reasons outlined in this chapter. In the next chapter, we'll discuss an alternative model and its evidence. Some archaeologists have suggested a hypothesized starting point of this migration: Japan.

The Out of Japan model rests primarily upon very striking similarities between Western stemmed points found at sites along the Pacific Coast and western interior of North America, and those found at sites throughout Japan and Northeast Asia. These Incipient Jomon sites date to between about 16,000 and 14,000 years ago. The people who lived at these sites hunted pigs, fish, dolphin, and turtles. They made bread from a flour composed of ground nuts and bird eggs, and nurtured the growth of wild plants, including beans and a kind of millet. These hunter-gatherers also made some of the earliest known pottery in the world after about 15,000 years ago, which was decorated with cord patterns (called Jomon, the origin of their name).

At the end of the LGM, proponents suggest, Incipient Jomon

hunter-gatherers moved northward into Northeast Asia. From there, they spread eastward along the southern coast of Beringia and continued along the west coast of Alaska of the Pacific Northwest, dispersing into the Americas along the "kelp highway" provided by marine resources.

BIOLOGICAL DISTANCE STUDIES

Before assessments of biological relationships between ancient individuals were made possible through DNA analysis, biological anthropologists relied exclusively on comparisons of morphological traits mostly from bones or teeth to answer questions about population history and biological relatedness. These comparisons assumed that certain physical traits had an underlying genetic basis, and that any two individuals (or groups) with similar physical traits were more likely to be related than those with dissimilar traits. This approach is called biological distance analysis (or often biodistance studies for short).

Biodistance researchers study metric traits (those with continuous measurements like length, breadth, and height) and discrete traits (those that are present or absent in a person, such as an extra suture in the skull).

Human bodies are remarkably variable, and this variation is thought to be under strong genetic influence. Thus, biodistance studies have been used widely in biological anthropology and forensic science, and analytical methods have become increasingly sophisticated. However, there are important caveats to their use.

In practice, the effects of the "environment"—a term that loosely describes anything nongenetic, such as the influence of nutrition, stress, disease, childhood development—almost certainly influence these traits. Thus, biodistance studies should be interpreted cautiously. The so-called Paleoamerican morphology, which we discuss in chapter 9, is an example of a hypothesis derived from biodistance studies, which was falsified by genetics research.

Ancient DNA is a more precise and reliable way of inferring biological relationships. However, because of the scarcity of ancient genomes, much of the work of understanding relationships between ancient groups continues to rely upon the study of morphology.

Researchers in this field have sought to circumvent these limitations by relying on traits that have been demonstrated to be strongly influenced by genes, such as morphological tooth traits, which have been shown in genetic analyses to be highly heritable. Researchers continue to develop appropriate methods for the study of such traits, many of which take into account underlying genetic and environmental influences (31).

However, the Out of Japan model isn't supported by biological evidence. Biodistance studies of dental traits (see sidebar: "Biological Distance Studies") show that the Jomon are unlikely to be ancestors of Native Americans, and this is supported further by studies of the genomes of both Jomon and First Peoples. Perhaps cultural diffusion—the spread of ideas and technologies—might be a better explanation for the striking similarities between Jomon and Western stemmed points (32).

What about the Really Old Sites?

The possibility that humans might have been present in the Americas at a very early period—between 200,000 and 50,000 years ago—has long excited archaeologists. As we discussed earlier in this chapter, before the invention of radiocarbon dating methods—or even the development of organized chronologies based on geological context—many claims were made about the "glacial man in America"(33). These claims were denigrated by scholars like William Henry Holmes and Aleš Hrdlička.

Perhaps the most famous example of a claim for a very ancient human presence in the Americas is the Calico Early Man Site. Located in the Mojave Desert, California, this site contains very clear evidence of human activities dating to an early period (10,000 years ago). But excavations by archaeologist Ruth Simpson turned up artifacts that she believed were much older—in geological layers that might have been deposited anywhere between 200,000 and 50,000 years ago. She brought in the famous paleontologist Louis Leakey to help her study the site. Leakey, an expert on early hominins and their stone tool industries, confirmed that many of the broken stones at the site were deliberately manufactured tools based on their resemblance to what he was seeing in Olduvai Gorge. However, Simpson and Leakey's claims did not stand up to scrutiny. The rocks were found in the middle of a geological deposit produced by a fast-moving river; their breaks could easily have been caused by the actions of water. Without any human skeletons or more positive evidence of human activity, the Calico Early Man Site's earliest dates are rejected most archaeologists (34).

Such has been the fate of all American Paleolithic sites proposed thus far. And yet, new claims continue to be made for very early sites. In 2017, a group of researchers led by Steven Holen published a paper in the prestigious journal *Nature* claiming that a site in Southern California provides evidence of humans in the Americas 130,000 years ago (35).

A peopling of the Americas this early would have occurred before the major migration of anatomically modern humans—people like us—out of Africa about 100,000 years ago. Different kinds of humans had been living throughout Eurasia before then. *Homo erectus*, humans with smaller brains and very differently shaped skull from ours (probably reflecting differences in their diets), had evolved in Africa around 1.9 million years ago and were dispersing throughout Eurasia by around 1.8 million years ago. Neanderthals, humans with larger brains and sturdier bodies than us, evolved from *H. erectus* populations in Europe around 140,000 years ago and spread throughout Eurasia. Denisovans, whose physical features are currently unknown to us because most of our information about them comes from their genomes, evolved from a common ancestor shared with Neanderthals around 450,000 years ago (probably *H. erectus*) and also lived throughout Eurasia.

Anatomically modern humans start showing up in Africa's fossil record as early as 300,000 to 270,000 years ago. They evolved gradually and also moved around a bit before their major migration out of Africa; some have been found in China dating to as old as 120,000 years ago, in Israel between 194,000 and 177,000 years ago, and in Greece as early as 210,000 years ago (36).

At this point in reading this book, you have probably already figured out that it can be difficult to reconcile different lines of evidence in understanding past events. This is true of human origins as well. On the one hand, we have fossil evidence, combined with genetic evidence in Neanderthal genomes of African *H. sapiens* populations reproducing with European Neanderthals sometime prior to 200,000 years ago. This seems to clearly point to a very early human movement out of Africa. But on the other hand, when you look at genomes from all ancient and contemporary *H. sapiens* populations, you see clear evidence of their migration out of Africa much more recently: only 100,000 to 60,000 years ago (37). There are several possible explanations (perhaps the Greek fossils were actually Neanderthals?), but one of them, as geneticist Aylwyn Scally notes, "is multiple or semi-continuous migration(s) of humans out of Africa" (38).

So could one of these possible early human dispersals have occurred in the direction of Siberia, across the Bering Land Bridge, and down through North America, all the way to California? Or could another kind of human, like a Neanderthal, *H. erectus*, or Denisovan, have gotten to the Americas? Most archaeologists and geneticists are very skeptical of either possibility. From an archaeological perspective, no skeletal remains that look even remotely like an early human have been found in the Americas, and we have none dating from anywhere near that age. We also don't see any unambiguous stone tools known to have been made by these earlier peoples: no Acheulian handaxes made by *H. erectus*, or Mousterian knives and scrapers made by Neanderthals, or Aurignacian blades made by Upper Paleolithic modern humans and Neanderthals.

What *is* found at the very early sites in the Americas tends to be either less direct evidence of a human presence or stones that some archaeologists suggest were flaked by humans. During much of early human history, many kinds of stone tools were created by flaking pieces off of a stone called a core; both the flakes and the core were used as tools. But how do you tell the difference between a flake removed from a larger stone by human hands (an artifact) and one that broke off naturally (a geofact)? Every professional archaeologist I've ever met has had to deal with countless numbers of people bringing them random rocks that they insist are tools.* Humans are extremely good at detecting patterns, even where there are none, and the field of pseudoarchaeology rests upon countless "evidence" for its non-mainstream claims from this phenomenon.

The validity of any early archaeological site in the Americas is assessed according to a specific set of standards. For a site to be accepted as genuine, it has to have securely dated, undisputable evidence of a human presence in an undisturbed stratigraphic context (39). In other words, the site has to have some object clearly created or used by humans (or human remains themselves), dated, in a geological layer that has not been disturbed or mixed with other layers.

Pleistocene environments constrain when people could have entered the Americas in the past. While it was at its full extent, between about 26,000 to 19,000 years ago, glacial ice would have prevented travel out of Beringia. Barring the possibility that there was some as-yet-undiscovered path through the ice wall, people could only have migrated *before* 26,000 years ago or *after* 19,000 years ago. Currently the weight of evidence leans more toward the post-19,000-year side, but there are some sites

---

* Despite the fact that I'm emphatically *not* an archaeologist, I have someone who periodically emails me photographs of rocks that they claim to be evidence of Europeans in the Americas predating the Clovis period.

in South America that have been recently discovered dated to as old as 30,000 years ago, which a couple of my archaeological colleagues find quite convincing. The footprints at the White Sands Locality 2 site provide even stronger evidence (if their 23,000-to-21,000-year dates are accurate) that at least some people were in North America during the LGM. Depending on how early their initial migration was, the interior route may have been a plausible pathway for them (40). For the most part, however, the majority of my colleagues are skeptical of sites this old. We will discuss genetics evidence later, but quite a few geneticists (I will go on the record here as one of them) are very much open to the possibility that people could have been in the Americas during or even prior to the LGM; we optimistically await the crucial evidence that could reconcile the archaeological and genetic records. We will discuss some potential models for reconciling the archaeological and genetic records more in chapter 7.

For early sites the key problem tends to be the need for an "undisputable evidence of a human presence." Is a rock flake a tool, or was it created naturally? Many things can cause a stone to break, including normal geological processes like water (as in the Calico Early Man Site case). Even the actions of animals like capuchin monkeys can result in the creation of broken rocks that could be mistaken for human tools (41).

Very early sites (especially those dating to before 50,000 years ago) need to be assessed with the same rigor applied to any other archaeological claim, and many of them fall short of the evidence needed to be acceptable, even to extremely open-minded archaeologists. Let's take a look at the Cerutti Mastodon site as an example. The claim it advanced was audacious, and it was published in what is arguably the most prominent journal in science. This site contained several rocks among the remains of many large Pleistocene animals, including those of a young mastodon whose bones had been damaged. The authors claimed that the damage on the bones was caused intentionally, and that the rocks had been used as hammerstones and anvils, pointing to a human presence at the site 137,000 years ago.

The majority of the archaeological community has found many problems with the evidence the authors presented as revealing a human presence at the site. The authors claim that the only possible explanation for the damage caused to mammoth bones found at the site was from humans smashing them in order to extract nutritious marrow from inside. But the experiments that they performed on elephant bones to demonstrate this were criticized as methodologically flawed and not up to the acceptable standards of the field of experimental archaeology (42).

Other archaeologists disputed the claims that the "hammerstones" could only have been made and brought to the site by humans. Still others argued that there were many other possible explanations for how the mammoth bones were fractured—including the fact that the site was near (and partially underneath) an area of road construction where heavy machinery was used prior to its excavation.

The last major line of critique points toward the importance of the site's stratigraphy. Archaeologists can't directly date every single object that they excavate from a site. In order to reliably associate any artifact with a particular date obtained from something else in the same stratigraphic layer, you have to be certain that the layers haven't been disturbed or the artifacts moved around or mixed up. Archaeologists have set conventions for reporting information from sites so that any other archaeologist can evaluate their evidence. Many archaeologists find the site stratigraphy reporting in the Cerutti paper to be inadequate, mainly because detailed maps of the stratigraphic relationships have not been provided by the authors (43).

Holen and his colleagues have argued spiritedly against every critique (44), but at this point, the vast majority of the archaeological community simply does not find these arguments persuasive on technical grounds (45).

Cerutti is just an example of how very early sites are evaluated by the archaeological community. Some people, pointing to the fall of the Clovis First paradigm, believe that strongly expressed criticisms by mainstream American archaeologists are *in themselves* a sign that a site is legitimate (46).

This is nonsense. Every scientific claim must stand or fall upon its own merits, and if the evidence supporting a claim falls short of acceptable standards, it is very appropriate—in fact it is *necessary*—to critique it.

Besides evaluating archaeological data used to support a claim about an early human presence in the Americas, it's also important to examine—with caution—how that claim fits with evidence from other fields. For example, current genetic models do not reflect an independent evolution of people on the American continents from archaic (*H. erectus*, Neanderthals, Denisovans, or an as yet unknown hominin) humans. Genomes from Indigenous peoples of the Americas and their ancestors show an unambiguous signal of descent from Upper Paleolithic populations of anatomically modern humans in Siberia/East Asia, with only the same traces of ancestry from archaic humans that are present in other populations throughout Eurasia. Genetic evidence does not fit with the paradigm proposed by the authors of the Cerutti paper (47). But these sites leave us with a testable hypothesis: If an early site like Cerutti is valid, then eventually other sites of similar age will be found. So far they have not.

New Models to Consider

If the evidence for archaic humans is lacking at present, does this mean that people weren't in the Americas prior to 16,000 or 15,000 years ago? Not necessarily. The key question is: *Which* pre-Clovis sites are legitimate evidence of a human presence in the Americas? And could there be earlier ones that we haven't yet detected? We have so many gaps in both the genetic and archaeological records that more data in the future may reveal additional evidence consistent with people here during or even before the LGM.

The White Sands Locality 2 site is a good illustration of this. At the time of this writing, the paper describing the site is just days away from publication (the authors kindly let me have an advance copy). By the time you are reading this book, the archaeological community will have extensively scrutinized and debated the evidence. Some will agree that this is a paradigm-changing site: concrete evidence of people in North America during the LGM. Others will find technical faults in the dates or stratigraphy of the site. They will have more trouble dismissing the site as lacking evidence of humans, as the footprints are spectacular and unmistakable.

Where are other sites like this? critics may ask. After all, New Mexico isn't exactly neglected by archaeologists. Shouldn't we have found other early sites by now?

The counterargument may be that there were so few people during that period that detecting their presence may be extremely difficult without a better understanding of how and where they lived. Perhaps, like the Folsom site, White Sands's enduring legacy is what it teaches us about where we should focus archaeological scrutiny.

It's also possible that there were people in the Americas at a very early date who did not contribute ancestry to any later (or contemporary) groups. We don't know how often this has occurred in human history, but examples have been identified through the study of ancient genomes. For example, the 45,000-year-old human from the Ust'-Ishim site in western Siberia, and the so-called "Ancient Beringians" from Alaska (who we will discuss later in this book) left no direct genetic ancestors that we can find (48). And we have archaeologically documented evidence of Norse presence in the Americas prior to 1490s, but as mentioned previously no genetic traces have been found from that group in the genomes of any ancient or present-day Native Americans.

In short, genetic evidence argues against the presence of archaic humans in the Americas during the Paleolithic. Archaeological evidence from extremely early sites such as Cerutti is unconvincing. Archaeological evidence from sites like White Sands that date to the LGM is much harder to dismiss. It may be that people were in the Americas by 25,000 years ago. Figuring out how that finding fits with genetic evidence is a new puzzle to solve. But I think that while we exercise a healthy skepticism, we must learn our lessons from the erroneous Clovis First paradigm and not simply dismiss evidence because it doesn't fit with a model we happen to favor (49)—we shouldn't simply replace Clovis First with a similarly dogmatic upper limit. All scientists must hold themselves open to the possibility that we could be wrong, and it may very well be that in 5, 10, or 20 years, this book will be as out of date as any other. *That* possibility is what makes working in this field so rewarding.

Chapter 3

Imagine living in eastern Beringia during the Pleistocene.

You and your band, a group of several extended families, follow a way of life that your ancestors did for many generations. It is summer, and you make your home close to the river, where you can hunt waterfowl and fish and feast on the berries and other delicious plants that grow in the valley. The hunters in your family spend their days stockpiling precious stone from nearby outcroppings, knowing it will be much harder to find good tool stone at your winter home, many weeks' travel from here. Late afternoons around the camp are punctuated by the sharp report of rocks striking each other as the hunters (and the children who imitate them) shape the rocks they've collected into lighter, more portable forms that they can use later in the year to quickly replace the tiny blades on knives and spears. One of the hunters is taking time off to recover from childbirth (see "Gender in Hunter-Gatherer Societies" sidebar); a young teenager who has gotten good at making tools has been selected to replace her in this essential task. This afternoon, he sits proudly with the other toolmakers, the envy of the younger children who redouble their clumsier efforts to work the less valuable practice stones. Laughter and songs fill the air; the barking of the dogs at some imaginary threat and good-natured arguments about who the newborn most resembles compete with the squabbles of the young toolmakers over who gets to sit closest to the hunters and watch them. The children are excitedly planning an expedition to see the ice wall, several days' walk from your camp. Some of the older members of the group have agreed to shepherd them; it will be an important chance to teach them and give their parents a bit of a break.

If the ancestors of the First Peoples came from Asia, then archaeologists believe that their paths must have gone through Alaska. And yet, despite intense interest in the area, the early archaeological record of the region does not lend itself to straightforward interpretation by geneticists. Or, as one of my colleagues put it more bluntly: "The early archaeology of Alaska gives me a headache."

Alaska serves as geographic bookends for the histories that make up the peopling of the Americas. This region played important roles during both the very earliest and the very final stages of the peopling process. In this chapter, we will take a close look at the first bookend, and how the archaeological record of Alaska leads archaeologists to wildly different interpretations of how the Americas were initially peopled.

During the LGM, most of the Arctic, including Western Siberia, Scandinavia, Greenland, nearly all of Canada, the Aleutian Islands, the Alaska Peninsula, and southeastern Alaska, was covered by glacial ice. The distribution of this glacial ice influenced where people went during the end of the Pleistocene. They could no more easily have gone into Canada and Greenland than they could have gone south; the way was blocked by ice. Coupled with the extreme environments of these regions, this meant that much of the lands north of 66° 32′ N (the Arctic Circle) weren't populated until after the rest of the Americas. People were living throughout lowland Alaska by at least 14,000 years ago but did not reach the Aleutian Islands until 9,000 years ago or the coastal and interior regions of Canada and Greenland until about 5,000 years ago.*

But there was one region of the Arctic that remained unglaciated throughout the LGM. Eastern Beringia—present-day Alaska—was an ice-free cul-de-sac at the end of the Bering Land Bridge (1). People could have lived there. *Whether* they did so is a question that fascinates many archaeologists.

GENDER IN HUNTER-GATHERER SOCIETIES

Historically, archaeologists have frequently used ethnographic analogy (the study of present-day cultures to understand the past) to interpret the archaeological record of the Americas. This approach led to a prevailing assumption that at most sites the hunting was done by men and the gathering was done by women. But over the past several decades, there has been a growing body of archaeological scholarship pointing out that this and other assumptions partially reflect contemporary Western interpretations of gendered activities and, to some extent, the people who have been doing the interpretations.

Understanding gender in past societies is complex. Among contemporary hunter-gatherer societies, the majority of hunting is performed by males. But assuming that

---

* We will discuss how they migrated into and adapted to these regions in later chapters.

this was also true of ancient groups is complicated by the fact that burials of women, as well as men, contain hunting implements in them, such as projectile points. For example, in 2013 the 9,000-year-old burial of a 17- to 19-year-old woman (whose sex was identified via both morphological and molecular evidence) was discovered at the archaeological site of Wilamaya Patjxa in Peru. Buried with her was a complete toolkit for hunting and processing big game, including projectile points, a knife and flakes for field dressing, and scrapers and choppers for processing hides and extracting bone marrow. Another individual buried nearby, identified as biologically male, had similar projectile points—though not the full toolkit—buried with him. If we can interpret the artifacts buried with these individuals as indicative of their activities during life, then we might reasonably conclude that both individuals were big game hunters, regardless of biological sex.

Notice that I'm careful to use the phrase *biological sex* here. Sex and gender are different things, although they are often confused by people who use the words interchangeably. In my field (biological anthropology), many scholars define sex in terms of physical differences: reproductive anatomy, secondary sexual characteristics, chromosomes. While a discussion of this topic is outside the scope of this book, it's important to note that there are no neat divisions between physically or genetically "male" and "female" individuals: Some people have reproductive anatomies that do not fall within this dichotomy, and there are a wide variety of chromosomal combinations (and associated physical attributes) beyond XY = male and XX = female. Biology is more complicated than that.

Gender, in anthropology, refers to both a person's internal identity and the socially constructed roles that people practice. Gender and sex may be aligned, or they may not. Many societies in the past and present recognize multiple genders beyond men and women and multiple ways in which gender is defined. We can't confidently assume that we know what a person's gender was simply because we can determine their biological sex from their DNA or the shape of their pelvis.

We don't know whether the first Wilamaya Patjxa individual was considered a woman by herself and others. Although she was biologically female according to her skeletal features and DNA, that may not have been her gender identity. Contemporary and historical Indigenous groups of the Americas—as in other societies around the world—have diverse conceptions of gender that don't necessarily align with the male/female duality imposed by Christian colonizers. (The same caveat applies to the second individual at the site—how do we know that he was considered to be a man by others or himself?)

A second complication lies in how to identify a person's role or status in life. Often, archaeologists identify someone's "profession" by the objects he or she was buried with: If they were buried with spears, they must have been a hunter or warrior. If they were buried with sewing needles, they must have been a tailor. If they were buried with certain sacred objects, they must have been a priest or shaman or holy person. If there were buried with exotic or expensive objects, they must have been an elite person or ruler, and so forth. But this approach can be misleading. People put all kinds of objects into the graves of their beloved relatives, not necessarily those that the person used in their lifetime. A person may have been buried with an array of items to provision themselves in their afterlife, and not all of them would have necessarily reflected that person's role in life. For example, two infants at the Upward Sun River site, who have been genetically sexed as female, and the toddler at the Anzick site, who was genetically sexed as male, were buried with spears. These children were not physically spearing and butchering Pleistocene mammoths. So the inclusion of spears in their burials must have had some other meaning. Perhaps they were intended to become warriors or hunters as adults, and their status was ascribed rather than earned. Perhaps their kin had an understanding of the afterlife as one in which the children would have used these objects, or they served as symbolic or sacred objects. We must be cautious in our interpretations. Regardless, the interpretation of one burial with hunting implements as a big-game hunter because he was biologically male and another buried with the same objects as not a big-game hunter simply because she was biologically female is unsound. It's very important to be mindful of our own biases in interpreting the archaeological record.

To get a broader perspective on just how frequently biologically female individuals may have been big-game hunters, the authors who reported the Wilamaya Patjxa burials did a systematic study of the association between individuals with inferred sex and the artifacts they were buried with at sites throughout the Americas. Out of 27 individuals buried in late Pleistocene and early Holocene sites with toolkits for hunting big game that were possible to sex morphologically, 11 have been sexed as physically female. This suggests, at minimum, that hunting may not have been viewed as an exclusively male activity across all time and localities throughout the Americas (2).

Gateway to the American Continents

If people migrated south of the ice sheets 17,000 to 15,000 or even 30,000 to 25,000 years ago, as we discussed at the end of the last chapter, we might expect to find archaeological evidence of humans in Alaska (eastern Beringia) during the LGM.*

We have not found this evidence. Or to put it a different way: we have not *yet* found *sufficiently convincing* evidence. We have

---

* Paleoecologist Scott Elias suggested to me another reason for why we might not see much evidence for people in Alaska during the LGM. The LGM was a period of extreme aridity across much of the planet. In Eastern Beringia, he told me, "virtually all the lakes dried up. This made Eastern Beringia a lousy place to live, as plants, animals, and people all need water. If they were like modern elephants, then mammoths needed about 700 to 1,000 liters (18 to 26 gallons) of water daily." People may well have been isolating in Beringia, but not spending much time in Eastern Beringia, leaving very few archaeological traces.

solid archaeological evidence of people living there after the
LGM, during the Late Glacial period (~14,000 to 12,000 years
ago), but claims of earlier human habitation are not accepted by
the majority of archaeologists, and even potential candidate sites
are very rare.

Some archaeologists are comfortable with the assumption
that there are gaps in the currently known archaeological record
of Alaska. In other words, they believe that the earliest (well-
accepted) archaeological evidence of people in Alaska, which
dates to about 14,200 years ago, is much younger than the actual
date when people first arrived. They claim that the archaeological record of Alaska is biased in two ways: First, much of central Beringia (the Bering Land Bridge) and the coastal regions
that would have been above sea level during the LGM are now
underwater and inaccessible. Second, because so much of it is
remote and difficult for excavation teams to access, only a tiny
fraction of present-day Alaska has been studied archaeologically.
Enormous expanses of land have yet to be surveyed for archaeological sites. Therefore, they argue, while one cannot assume
people were there based on the traces of indirect evidence found
thus far, it's far too early to definitively say anything about when
people were first in eastern Beringia, especially considering that
the first populations would likely have been small and dispersed,
leaving a very light archaeological footprint (3). In other words,
we are looking for very small needles in a very large haystack.

But other archaeologists are uncomfortable with that assumption. Instead of hypothesizing about sites that have yet to be
found, they argue, we should look at the story the existing
archaeological record of Alaska tells us.

They believe that the archaeological record within Alaska—
particularly the lithics present at late Pleistocene sites—supports

a later peopling model. They interpret a set of archaeological sites in interior Alaska around the middle Tanana and Nenana Valleys—the earliest undisputed sites in Alaska—as the traces of the earliest human presence in the Americas. In this chapter, we will explore this alternative model, the evidence it rests upon, and the assumptions it makes.

THE IMPORTANCE OF LITHICS

Lost and broken stone tools give us a very intimate glimpse into the lives of ancient peoples. From the type of tools and the wear upon them, we can discern what activities they were engaged in and the strategic choices they made in how to survive environmental challenges. The materials used to make the tools tell us what was available to them, and how far they ranged or traded to obtain them. We can learn about the skill level of the individual toolmaker by the way the tool was manufactured. We can learn about how an individual site was used by the distribution of stone tools and the debris created from their manufacture. We may infer the seasonality of site use from the distribution of stone tools across sites, and (with caution) information about the existence of different social groups and their territorial boundaries.

Archaeologists are sometimes criticized for their fixation on stone tools as cultural markers at the expense of a more thorough examination of other aspects of material culture—and the people who produced them (4). But there's also the reality that lithics (stone artifacts) are the most likely things to be preserved in the archaeological

record because of their durability. This differential preservation biases the archaeological record, and therefore any
inferences about culture, identity, and lifeways from lithics
alone should be treated cautiously.

Late Pleistocene Archaeology of Alaska

Recall that in the early decades of the 20th century, figuring out
the chronologies of sites was a major challenge for archaeologists;
they had to rely upon *relative* dating methods in reconstructing
ancient history. In much the same way as physical anthropologists constructed racial categories, archaeologists classified artifacts into types according to their physical characteristics (shape,
material, method of creation). These attributes were widely
believed to be at least partially reflective of the cultural identity
of the people who made them. Groups of particular artifacts reliably associated with each other in a limited geographic region or
period of time were organized into *complexes,* often used as prox-
ies for groups or populations. Groups of particular artifact styles
that persisted over great periods of time and were geographically
widely dispersed (indicating a sustained cultural identity or technological approach) were referred to as *traditions.* Transitions
between traditions have often been identified by major technological changes in the archaeological record.

Archaeologists sorted artifacts into relative chronological order by categorizing artifacts and looking for associations
between them and markers of age (such as stratigraphic position
or co-occurrence with the remains of an animal known to have

gone extinct by a certain period of time). Prior to the invention of radiometric dating techniques, archaeologists figured out approximately when people were present at a site based on the artifacts they left behind.

This is how fluted projectile points became the marker for the early peoples of the Americas, as we discussed in the last chapter. Archaeologists believed that because Alaska must have been the gateway through which people must have initially entered the Americas, they would find the precursors to fluted projectile points—if not the points themselves—in layers predating Clovis at sites just north of the entrance to the ice-free corridor. Blufftops overlooking the Tanana and Nenana river valleys in this region yielded dozens of sites that archaeologists confidently dated (once radiocarbon methods were available) to the late Pleistocene.

But what they found at these sites was not what they expected.

Excavations of the deepest layers of sites in interior Alaska repeatedly turned up complex and diverse arrays of stone artifacts. The oldest (undisputed) evidence of humans in Alaska is currently from the Swan Point site, in the Tanana River Valley. Around the remains of ancient campfires, archaeologists have found a scattering of stone flakes, tools, and debris from processed mammoth ivory, antler, bone, and stone in layers dating to about 14,200 years ago (referred to as Cultural Zone 4b or CZ4b). The artifacts found in CZ4b seem to reflect a very brief, single event in which a group of people lived at the site for a number of days or weeks. Archaeologists generally interpret this evidence as indicating that CZ4b was a short-use hunting camp. The tools that people manufactured at Swan Point CZ4b included a particular set of lithics: tiny microblades and the tools used in their manufacture (see "The Microblade Toolkit and How It Was Made" sidebar) (5).

Dyuktai complex microblade core from the Ushki-5 site. Microblades would have been struck from this prepared core. Redrawn from a 2003 article by Ted Goebel in *Science*.

Denali complex microblade core from Donnelly Ridge. Redrawn from a 2017 *Current Anthropology* article by Kelly Graf and Ian Buvit.

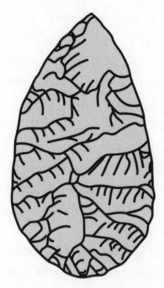

A teardrop-shaped Chindadn point, which is associated with the Nenana complex. Redrawn from a 2001 *Arctic Anthropology* article by John F. Hoffecker.

Mesa complex projectile point from the Mesa site. Redrawn from a 2008 *Journal of Field Archaeology* article by Michael Bever.

Between about 13,500 and 12,800 years ago, at numerous sites in interior Alaska (6), archaeologists found quite different artifacts: no microblades, but stone tools that had been flaked on both sides (bifacially flaked), including small teardrop-shaped spearheads called Chindadn points. They designated this group of artifacts the Nenana complex.

Numerous sites dated to a bit later in the Tanana and Nenana river valleys are marked by microblades, microcores, and associated tools (burins and end scrapers). These are referred to as Denali complex sites, and they date to between approximately 12,000 and 6,000 years ago.

A final group of sites in the Brooks Range, northern, and western Alaska yielded bifacially flaked oval points that tapered on each side. Like Clovis points, these resembled lance heads, and are called "lanceolate" bifaces. Archaeologists have classified sites with these Clovis-like points as belonging to the Northern Fluted, Mesa, and Sluiceway complexes (7).

Stone Tool Industries of Beringia

Complex	Tools	Dates	Technological links
Dyuktai	Yubetsu-style wedge shaped microblade cores	14,200 years ago (Swan Point)	Dyuktai sites in Siberia
Denali	Microblades, Campus-type wedge-shaped microblade cores, burins, end scrapers, osseous projectile points with slots for blades, lanceolate-shaped bifacial points	12,000–6,000 years ago	Possibly descended from Siberian Dyuktai complex, though made using a different method

Nenana	Chindadn points, no microblades (or associated artifacts), end scrapers, bifaces	13,500–12,700 years ago	Chindadn points first seen in western Siberia at the Berelekh site (14,900–13,700 years ago) and Nakita Lake (13,700 years ago), and at Little John (14,050–13,720 years ago)
Mesa	Bifacial lanceolate projectile points	<13,000 years ago	

The diversity of the archaeological record in this region lends itself to several interpretations. The oldest site in Alaska, Swan Point CZ4b, shows us that the microblade toolkit is the oldest stone tool industry in Alaska. But the microblades at Swan Point CZ4b were made in a different way than Denali microblades (see "The Microblade Toolkit and How It Was Made" sidebar). Chronologically, there is a long period of time between CZ4b and the appearance of Denali microblades; during this period of time, people were making tools characteristic of the Nenana complex.

It can be tricky to equate a toolkit with a culture, in Alaska or anywhere else. Nevertheless, this is a starting point for two different interpretations of the archaeological record (that can perhaps be tested with genetic evidence). Some archaeologists argue that different toolkits across Alaska were made by individual groups of people with diverse cultural/technological strategies for different regions and/or seasons. For example, microblade-based tools might have been preferentially used during the winter, when access to stone sources was limited because of snow or ice.

As envisioned by some archaeologists, a single broad tradition, referred to as the Paleo-Arctic or Beringian tradition, would have extended across the entirety of Beringia—a truly gigantic geographic territory (8).

THE MICROBLADE TOOLKIT AND HOW IT WAS MADE

Stone points were often lost or broken in the course of hunting, as archaeologists have discovered many times. Replacing them can be relatively straightforward for a competent toolmaker if there's good quality stone readily available. But what about during the winter, when rock deposits are buried under snow or ice? What if you are many miles away from sources of good rock?

The microblade toolkit was a technological response of men and women who faced unique challenges to hunting in an Arctic environment. In contrast to lanceolate (lancelike) projectile points that were fashioned out of a single piece of stone (like Clovis points), microblade-based tools used dozens of tiny stone blades inserted into osseous (bone or antler) tools to create a composite cutting edge.

The microblades were struck off of cores, small rocks that had been carefully flaked in order to yield blades of a consistent size and shape. Burins, carefully shaped stones, were probably used in the preparation of these microblade compound tools. End scrapers were fashioned by removing flakes from one end of a core. The sharpened surface may have had a variety of uses, including processing hides from animals and shaping antler, bone, or wooden tools. Bifaces, or stone knives, axes, or points flaked in such a way as to produce two sides, were common components of many different kinds of toolkits, including microblade assemblages.

Toolmakers prepared numerous cores (microcores) in advance, possibly when staying at encampments such as Swan Point CZ4b. They would then carry these cores with them as they traveled, ensuring that they had the ability to make tools even when far away from the sources of good-quality stone. If their spearpoints were damaged in the course of hunting, it would have been relatively easy and economical to replace lost or broken microblades by flaking them off of their pre-prepared cores and fitting them into the bone projectile point. A skilled toolmaker could certainly replace projectile points created out of a single piece of stone, but it would have been more time consuming, more costly in terms of raw materials, and more cumbersome to carry around the larger cores required to make bifacial projectile points than microblades.

Microblades appear at sites in interior Alaska for over 10,000 years, indicating their success as an adaptation. But they weren't a static technology; we know from the archaeological record that people made these tools in at least two major ways. Toolmakers using the Yubetsu method prepared a bifacially flaked, leaf-shaped blank.

They would then strike the biface along its top ridge to create a flat platform and carefully shape the angle of one end. Then they removed tiny flakes from one end by applying pressure with a bone or antler.

Though the exact location and timing of its origin is disputed, the Yubetsu method came to be used by peoples widely dispersed across Beringia and Asia, including people from the Dyuktai culture of Siberia. Yubetsu is seen at Swan Point CZ4b but disappears from the Alaskan archaeological record after that.

The Campus method seems to have evolved from the Yubetsu method. At Swan Point and other sites, people were making microblades using the Campus technique beginning around 12,500 years ago.

Toolmakers using the Campus method started by preparing a blank from a flake. They would shape the flake along one side. They then created a platform by hitting the top edge with lateral strikes. One end was shaped, and the microblades were removed using pressure flaking. The Campus method often required frequent reshaping (or rejuvenating) of the platform as microblades were removed.

The Campus method has so far only been observed in high frequencies at Denali complex sites and in Alaska and the Yukon. It is not seen in Siberian sites (although there are some similar examples in other parts of Asia). Because it can be used with stones of many different shapes and doesn't require bifacial flaking, it is more efficient and more economical than the Yubetsu method.

Why did methods for producing microblades change in

Alaska? This is an important question, but we don't have an easy answer to it. One explanation is that while tools belonging to the Nenana complex tended to be made using raw materials that were locally sourced, Denali complex tools were made from stone imported from more distant places. This hypothesis suggests that perhaps the changes in methods were adaptations for working with different kinds of stone. At Swan Point, the shift in microblade production methods is accompanied by a shift in diet; people began to eat bison and elk instead of mammoth and horses. This occurs shortly after the start of the Late Glacial interstadial, a period between 14,500 and 12,800 years ago in which the climate was much warmer and wetter than it had been previously. Perhaps the technological and dietary changes reflect some adaptations to the warming climate (9).

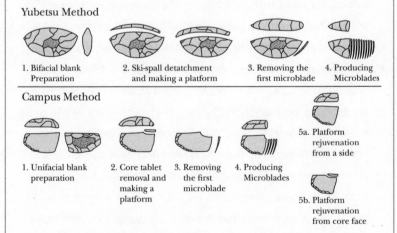

Yubetsu and Campus methods for preparing microblade cores and microblades. Redrawn from a 2017 *Quaternary International Volume* article by Yu Hirasawa and Charles Holmes and inspired by the 2012 book *The Emergence of Pressure Blade Making* by Yan Axel Gómez Coutouly.

It seems more sensible to other archaeologists to view the Denali, Nenana, Northern Fluted Point, Mesa, and Sluiceway complexes as representing distinct groups of people with different cultural and technological practices, occupying more or less the same region. The fact that these different toolkits don't co-occur at any sites (e.g., no wedge-shaped cores or microblades ever turn up in Nenana sites) is seen by these archaeologists as a significant argument against the pan-Beringian tradition (10).

EVIDENCE FOR CHILD TOOLMAKERS?

Throughout this book a number of genomes are discussed that were sequenced from the remains of children. Child mortality was a tragic but common occurrence throughout the past; in the absence of antibiotics and vaccines, infections were often deadly, and children were especially vulnerable during times of climatic hardship and limited resources. As a mother of a toddler, I find it excruciatingly difficult to write dispassionately about this subject.

Although children's remains are found all too often in cemeteries and isolated burials, their activities in life are surprisingly hard to detect in the early archaeological record of the Americas. In more recent archaeological periods in the Americas, a number of artifacts have been found that have been interpreted as toys: miniature pots, small projectile points that may have been used for child-sized bow-and-arrow sets. But we know almost nothing about the day-to-day lives of children in the late Pleistocene and early Holocene.

Archaeologists who study this period have not historically prioritized the study of children's culture. But recently there has been some research exploring this issue in the context of stone tool production, with some fascinating results.

The skill required to competently make—or *knap*—stone tools is not easily acquired I discussed the footprints at White Sands in chapter 2. But there are other traces of children in the archaeological record, including lithics. People who become skilled knappers do so only after many hours of observation, hands-on experience, and feedback from experienced toolmakers. Therefore, it seems reasonable that children who intended (or were expected) to become toolmakers must have begun learning the process fairly early.

Operating from this assumption, and from ethnographic studies of both children and college-aged students in flintknapping classes, what might a signature of beginning flintknappers (whatever their age) look like in the archaeological record?

Archaeologists grappling with this question suggest that the first place to start is looking for obvious mistakes in the process, particularly those that reflect poor motor coordination (which may be particularly associated with younger learners) and/or demonstrate a poor understanding of how rocks will fracture when struck in particular ways.

Recently, a team of researchers led by Y. A. Gómez Coutouly have articulated different mistakes one might expect to find at different stages of the learning process (which they suggest spans childhood through adolescence), translated such mistakes into specific expectations for what might be found in the archaeological record, and used this approach to look for evidence of apprentice knappers at two sites in Interior Alaska: Swan Point (CZ4b, dated to ~14,000 years ago) and Little Panguingue Creek sites (dated to ~10,000 years ago). These two sites have extensive debris from toolmaking activities, specifically the production of microblades. The archaeologists examined the shape of microblade cores and microblade core preforms to look for damage consistent with either the work of apprentice or skilled knappers. The researchers observed a number of examples of skilled knappers at both sites. They also identified the work of apprentice knappers, which seemed to span a range of abilities. They also found evidence at both Little Panguingue Creek and Swan Point sites of apprentices who clearly understood *theoretically* how to shape the tools but didn't seem to have the skill to carry out the process to completion.

At Little Panguingue Creek, which seems to have been a residential camp where people of all ages lived, rather than a short-term hunting camp like Swan Point, researchers found a stone flaked by one apprentice knapper who had a very poor understanding of the process and lacked the motor skills necessary to exert control over their striking. It's very tempting to imagine this stone was the product of a child trying to imitate adults or older children.

Interestingly, the authors observed that most of the cores hypothesized to have been made by apprentice knappers at both sites were clustered together, located around the margins of the tool production area at the site where skilled knappers were working. There's some evidence that the learners were using poorer-quality stone to practice on as well, reserving the high-quality toolstone for the experienced knappers (though they also shaped poorer-quality stone as well).

This research gives us a wonderful and fascinating glimpse into the day-to-day lives of younger people during this period. As the authors note in the last line of their paper, "No doubt these prehistoric boys and girls were frustrated at their knapping errors; but these were errors that it was right to make, for they led step by step towards the mastery of the skills required to their survival" (11).

Denali First?

Interpreting the relationship between Denali and Nenana toolkits (or more accurately, the people who made them) is sufficiently difficult. Figuring out how the early archaeological record of Alaska relates the earliest peoples south of the ice sheets is far more difficult.

We can start by ruling out one hypothesis. The lanceolate points of the Northern Fluted Point and Mesa complex (and Northern Paleoindian tradition more broadly) were originally thought to be a direct technological ancestor of Clovis, which would have provided a clear-cut evolutionary progression from Beringia to the Plains. But stylistic analyses have shown that the Northern Fluted Point complex appears to have been derived from (not ancestral to) projectile points made in the northern Great Plains. Furthermore, the Mesa sites date to 12,400 years ago—well after the appearance of Clovis at 13,500 years ago. Thus, archaeologists generally interpret these sites as having been created by people who had moved *northward* from the Plains around 12,000 years ago, rather than the ancestors of the first peoples below the ice sheets (12). Here is another example of how the movements of ancient peoples are always more complicated than our reconstructions would suggest.

To some archaeologists, the finding of 14,200-year-old Dyuktai complex microblades at Swan Point is the key to understanding the peopling of the Americas. As the Northern Hemisphere warmed after the LGM, people moved back into northeastern Asia between about 18,000 and 15,000 years ago. We see evidence of people living in Eastern Siberia at a group of sites in Kamchatka, Chukotka, Kolyma, Yakutia, and the Trans-Baikal regions, which archaeologists classify as belonging to the Dyuktai complex. The Dyuktai complex toolkits look very familiar: microblades and wedge-shaped microblade cores, burins, and scrapers, exactly the toolkit found at the Swan Point site.

Many archaeologists classify the wedge-shaped microblade cores found at the earliest layers of Swan Point as belonging to the Dyuktai complex themselves, a clear sign of cultural contact and/or migration across Beringia during the late Pleistocene.

To explain this, one group of archaeologists favors a model, first proposed by archaeologist Frederick Hadleigh West (13), which I'll refer to as "Denali First" for convenience's sake.

Migration and contact—in both directions—between Arctic peoples of Asia and North America has occurred frequently throughout history, and we'll discuss several examples in a later chapter. But the Denali First model suggests that Swan Point wasn't just a random incidence of people migrating from Siberia; it is a reflection of the earliest migration of Native American ancestors.

After all, proponents argue, what sites in Alaska are older? There are none that they find convincing. If we can assume that Dyuktai represents the toolkit used by the immediate ancestors of the First Peoples, then they would have migrated across the Bering Land Bridge from Siberia between 15,000 and 14,000 years ago.

A corollary to this model, advocated by some archaeologists, is that the first migration south of the ice sheets was down the ice-free corridor sometime between 14,000 and 13,500 years ago. As they migrated southward, these early groups of people would have abandoned microblade technologies and developed the fluted points that are hallmarks of the Clovis technocomplex. Under this model, Clovis was born directly out of the Beringian tradition, and the ice-free corridor was the most likely route for the ancestors of Clovis peoples south of the ice sheets. The earliest sites in interior Alaska are located just northwest of the ice-free corridor's entrance, positioning their inhabitants extremely conveniently for a southward migration as soon as it was possible.

Advocates of this model suggest that the reason earlier coastal sites haven't yet been found by archaeologists isn't because they were submerged by rising sea levels at the end of the LGM (as other archaeologists hypothesize), but rather because they don't exist. Further to this argument, peoples of the Dyuktai culture—the "mother culture" of the eastern Beringians—didn't use boats; they hunted large mammals (mammoths, horses) and lived inland. Archaeological sites in the Tanana basin don't have any evidence of maritime technology either. Thus, proponents argue, there's no reason to believe they would suddenly invent these adaptations in time to travel along the coast (14).

The model as I describe it above may seem like the simplest explanation for the origin of Native Americans. It doesn't require the assumption of earlier archaeological sites in Alaska, western

Beringia, or along the West Coast that either haven't been found yet or can't be found because they're under 100 meters of ocean. It doesn't require archaeologists to accept the validity of sites south of the ice sheets that don't look like what they would expect for the technological ancestors of Clovis.

It is a simple, elegant, and very testable model.

But it fails to explain all of the evidence outside of Alaska, particularly the presence of sites south of the ice sheet that pre-date Swan Point's 14,200-year-old occupation (15). In assuming that the Dyuktai complex is the best candidate for the material culture carried by the First Peoples, and that Swan Point is effectively the earliest evidence we have of these groups, it is essentially a new version of the Clovis First model. It requires a peopling of the Americas to have occurred very late (15,000 to 14,000 years ago, or perhaps even as late as 13,500 years ago), which is well after the earliest appearance of sites south of the ice sheets.

Proponents of the Denali First model have several explanations for the existence of pre-Clovis sites in the framework of their model. First, some assert that the vast majority—or possibly all—of pre-Clovis sites are not valid. That is, they fail to meet the standards of evidence for a human presence, or their dates are unreliable. This is robustly refuted by the archaeologists who have excavated these sites, who argue that the continuous skepticism about each and every pre-Clovis site is a perpetuation of the Clovis First mentality that has hindered progress in the field. *Even when our sites meet the standard of evidence that you demand, you still reject them because they don't conform to your cherished model,* they argue.

A second argument is that the artifacts found at pre-Clovis sites don't have any orderly and unified cultural and technological antecedents of Clovis. *But should they?* respond the critics.

We don't see this in Alaska either. Maybe the widespread, nearly homogeneous Clovis complex is important not because it represents the traces of the First Peoples in the Americas, but because that very uniformity itself is so unusual. *Would we expect small and mobile groups of people, widely dispersed across great distances, to have a tidy and cohesive technological evolution?* We don't see this in biological evolution; it's far more complex and messier. Why expect it in the archaeological record? Isn't it more likely that we should find diverse regional technologies, each adapted to their particular environments?

Multiple Dispersals?

Other archaeologists favor a different model: that the diverse tool traditions seen in Alaska represent multiple dispersals—perhaps two or three—of people out of Siberia into eastern Beringia.

One group of people used the Siberian Late Upper Paleolithic Diuktai complex toolkit; the traces of which appear at the Swan Point CZ4b. Another group used the Berelekh-Nenana complex toolkit. The later Denali complex may represent a third dispersal, or it may be a technological evolution from Diuktai. This model suggests the presence of multiple populations of humans in eastern Beringia during the late glacial (16).

Proponents of this model have multiple opinions as to whether Swan Point represents the earliest presence of people in Alaska.

Some proponents are highly skeptical of the assumption of an archaeologically invisible founding population. These archaeologists note that the 30,000-year-old archaeological sites in western Beringia known as the Yana Rhinoceros Horn Sites—which we will discuss more in Part III—show us the kind of archaeological evidence we might expect to find from a sizable population living year-round at a single location. They correctly point out that we see nothing like that before or during the LGM in eastern Beringia.

So why would populations have expanded northward and eastward during terrible climate conditions, when everywhere else in the world we see people retreating southward instead? they ask.

If people were in Beringia during the LGM, why haven't we found any traces of them? Don't construct archaeological models based on non-existent sites, they admonish.

Other archaeologists believe that multiple dispersal model is not incompatible with there being earlier populations in the region. This view rests upon evidence for pre-Clovis sites described in the previous chapter, as well as some more disputed evidence for a human presence in Alaska which we will discuss in chapter 6.

Another explanation for the presence of pre-Clovis groups below the ice sheets is that they were the result of "failed migrations," or dispersals into the Americas that did not contribute genetically or culturally to the First Peoples. From a genetics standpoint, it's perfectly reasonable to expect that not every population's genetic legacies would persist in a region over time. As discussed before, we see evidence of populations that have no known present-day descendants from the genomes of ancient people across the world, and it is certainly possible in the Americas. But to apply the term *failed migrations* to these cases is demeaning and highly problematic from an archaeological standpoint. " 'Failed migration' is a phrase used to sweep (pre-Clovis sites) under the rug and not confront or think about (them)," archaeologist Michael Waters told me in an email. They were people with their own histories and stories that deserve to be acknowledged as more than "failures," regardless of whether or not they contributed DNA to later generations.

Over the last two decades, developments in archaeology and genetics have forced researchers to look for new answers to old questions long thought resolved.

At present, it's difficult to find two archaeologists who agree on exactly how the Americas were peopled; they differ on which kinds of evidence they find most convincing in accepting or rejecting the validity of ancient sites, how different sites relate to one another, and how archaeological evidence should be integrated with genetic data (which we will discuss in the next chapter). Nevertheless, archaeologists' views today tend to cluster into a few general models.

Many archaeologists believe that people entered the Americas after the LGM as soon as a route was opened along the west coast of Alaska, perhaps as early as 17,000 to 16,000 years ago. A few others see the totality of the evidence, including the White Sands Locality 2 site, as supporting an even earlier migration, between 30,000 and 25,000 years ago. We can loosely group these archaeologists into the same category: those who agree on a pre-LGM peopling, but disagree as to the details.

Another model is based on sites like Cerutti, which point to a very early migration (137,000 years ago or earlier), probably by a different kind of human who was not ancestral to Native Americans. This "Paleolithic peopling model" is rejected by nearly all archaeologists.

Finally, as we discussed in this chapter, some researchers maintain skepticism about the validity of *all* pre-Clovis sites, maintaining that Clovis was first after all. This group of researchers believe that the archaeological record best supports a late peopling model, based upon clear cultural connections between stone tools found in Siberia and the stone tool technologies in eastern Beringia (present-day central Alaska). These archaeologists tend to be skeptical of the validity of the majority of pre-Clovis sites and instead favor an expansion of people from Siberia to Beringia between about 16,000 and 14,000 years ago, with a subsequent migration southward, probably down the ice-free corridor.

These models as I've presented them are not, of course, the only possible archaeological explanations for the peopling of the Americas held by researchers in the field. Which model—or which aspects of each model—archaeologists find most convincing depends on how they prioritize different kinds of evidence.

In 2018, I attended a talk by archaeologist Ted Goebel in which he succinctly summarized the major questions that researchers in the field are currently trying to address:

1. Who were the First Peoples of the Americas?
2. From where did they come?
3. What routes were taken?
4. When did the peopling occur?
5. How did dispersing populations move through the "empty" landscape?

These are the questions that geneticists have focused on in recent decades. In the coming chapters, we will explore the answers they have found thus far.

PART II

Chapter 4

Twenty years ago I took my shoes off and stepped into the Mayan underworld.

The darkness in the cave rendered my peripheral vision useless—the only objects I could see were those directly illuminated by the narrow cone of light from my headlamp, giving me only piecemeal images of the huge cavern. The archaeologists who were guiding us and supervising our training called this room the main chamber. Seeing a world through a headlamp is like looking through the cardboard tube at the center of a paper towel roll; you focus on one object, shift your head, take in something else, shift again. I felt like I was piecing together the entire cavernous chamber in a hundred spotlit gazes.

Against one wall was a cluster of beautiful white columns made up of fused stalactites and stalagmites. Over thousands of years, the drip of water from the limestone ceiling formed the flowstone deposits that glittered in the light of my headlamp.

As I turned my head slightly to the right of the columns, my light revealed something much younger resting against the wall: a manos and metate—stones the ancient Maya had used to grind corn.

I shined my headlight down to my feet, clad only in wet woolen socks. In order to protect both the artifacts and the cave formations, we had left our shoes outside, and now I was standing in the dark with the other students, my toes hanging off of the lip of one of countless small travertine dams that extended outward from the entrance of the main chamber to its back wall. Each dam was a ledge of calcite encircling small indentations in the ground, caused by the water that had once dripped from the countless stone icicles hanging from the ceiling.

The pools were dry now, but they hadn't been when the ancient Maya last visited this room, over a thousand years ago. In these pools, on the ledges where we now stood, and wedged into the crevices of the shimmering columns, the Maya had left dozens of pottery vessels.

There's something about seeing an object in situ* that helps you understand its purpose in a way that no information card in a museum can ever truly replicate. I was overwhelmed by these slivers of history I was seeing through my narrow light tube—I had never seen such a magnificent collection of pottery in one place outside of a museum. The sight of a gorgeously crafted pot sitting in one of the dried pools with a "kill hole"† drilled into its side and another pot with a fragment of its rim removed allowed me to imagine the person who left it there far more viscerally than any history text on the subject of ritual termination ceremonies. The objects were offerings, possibly a way to maintain relations with the forces concentrated within the cave.

---

* In situ (literally "in place") means an artifact is in the place where it was originally left; it is in its primary context or, in other words, undisturbed since the time it was deposited.

† A kill hole is a hole deliberately drilled into an artifact such as a pottery bowl, rendering its practical function useless. One interpretation of kill holes is that they were ritual in nature.

Ancestors were also left in this sacred space. The skeletons of 14 men, women, children, and infants had been left here, their bones glittering with the same mineral sheen as the formations on the ceiling and cave floor.

After spending my childhood in the Ozark Highlands exploring caves as junior member of a caving club, caves are comfortable places for me. I love their peculiar smell, like the freshly exposed earth after a spring-night thunderstorm, or that final waft of autumn that calls an end to camping season. It's that smell and the burst of cool air that hits you first at the entrance to a cave, signaling that it's time for the rituals of checking your light sources and extra batteries, and adjusting your gloves and kneepads. I love the moment when, after you've turned a corner or scrambled up some breakdown from the ceiling, you have passed beyond the twilight zone near the cave's entrance to deep and total darkness. Many people find the utter lack of natural light disorienting or frightening, a place where vaguely imagined fears (or real ones for people who are frightened of bats) begin to intrude on their consciousness. But I love it. The darkness and the thick silence allow you to hear every drop of water shaping the rock around you.

I was taught as a child that a cave is not a place of danger, as long as you respect it and follow the rules. Always have at least three sources of light, extra batteries, at least two reliable companions, and people outside who are aware of your plans and an idea of when to expect you back (with a cushion of an extra hour or two, in case there are new passages). Always wear a good helmet, gloves, and kneepads. Never touch a speleothem (cave formation), because the oil from your hands will damage or destroy it. Stay away from hibernation caves during the winter so you don't harm the bats. Advocate for conservation of caves and their inhabitants to anyone you meet, and carry out the trash you find that's been left by other people.

Take nothing but pictures. Leave nothing but footprints. Kill nothing but time. (And vandals, some of my father's friends added, darkly—the gallows humor of long-suffering veterans of cave conservation.)

And, I was told, never ever go cave diving (1).

In return for your respect, caves offer you the unique experience of seeing unparalleled treasures of nature: speleothems of the most astonishing beauty created over thousands of years by what began as a tiny accretion of minerals in water droplets. You have to move with utmost care to avoid touching them as you scramble over rocks or crawl through tunnels. Since your light source is usually a focused beam from a headlamp or flashlight, you learn to maintain a constant state of alertness in the underground world. As a child (and later as a teenager) I loved feeling this single-minded focus for hours, listening to the small sounds of water, our own footfalls, and the occasional flutter of bat wings, and glimpsing something ancient and beautiful in the beam of my flashlight every time I turned a corner.

Entering a sacred burial space like Actun Tunichil Muknal* requires an additional level of respect: for the place itself, for its history, for the ancestors interred there, and for the living people who still consider it sacred. It's an important consideration when visiting such places as tourists; one must be mindful of how the vocabulary of "discovery" and "adventure" and the opportunity to gawk at the remains of ancient peoples may be demeaning to them and harmful to their descendants.

---

* Actun Tunichil Muknal (ATM) is found near the present-day town of San Ignacio in the Cayo District of Belize.

To the ancient Maya, caves were sacred. Some contained entrances to Xibalba, the "place of fear," an underworld city below the surface of the earth ruled by the lords of death. According to the Popol Vuh, one of the few holy books of the Maya not destroyed by Spanish missionaries, Xibalba "is crowded with trials."

The first of these is the House of Darkness, where nothing but darkness exists within. The second is named Shivering House, for its interior is thick with frost. A howling wind clatters there. An icy wind whistles through its interior. The third is named Jaguar House, where there are nothing but jaguars inside. They bare their teeth, crowding one another, gnashing and snapping their teeth together. They are captive jaguars within the house. The fourth trial is named Bat House, for there are none but bats inside. In this house they squeak. They shriek as they fly about in the house, for they are captive bats and cannot come out. The fifth, then, is named Blade House, for there are only blades inside—row upon row of alternating blades that would clash and clatter there in the house (2).

Actun Tunichil Muknal (or ATM, as it's informally known) was one of the entrances to Xibalba. It seems to have been mainly used during what archaeologists call the Late Classic to early Postclassic periods, between 800 and 1000 CE. The people who visited this space would have had to swim through the deep blue spring-fed pool underneath the massive arched entrance into the cave. Moving deeper into the cave, past the twilight zone and into complete blackness, they would have scrambled up a steep climb (dicey even for people with helmets and headlamps) and ascended to a high passage that led to the main chamber.

There, in the presence of the stunningly beautiful formations, they left offerings of pottery, stone, jade, pyrite, and people (3).

Many of the people interred here had flattened foreheads, the result of deliberately shaping of the bones of a child's skull as they grew. Some individuals had their teeth modified by filing them. Like many societies, the ancient Maya sometimes employed aesthetic practices to make members of elite families more physically distinctive.

The nature of the "burials"* suggested that the people had been sacrificed (4).

After we spent time quietly (and carefully) moving through the main chamber in our stocking feet, the archaeologists guided us to an even more secluded sacred place, an alcove above the main passageway that has been named the Stelae Chamber. Within the Stelae Chamber, the ancient Maya had constructed two prominent monuments made of slate, which were propped up with broken cave formations. One of the monuments was scalloped on either side, giving it the appearance of a stingray spine. The other monument was carved so that it tapered at the apex, closely resembling a naturally occurring obsidian point. Both of these monuments have been interpreted by Jaime Awe, the distinguished Belizean archaeologist who has excavated Actun Tunichil Muknal and countless other sites throughout the region, as references to implements used in bloodletting ceremonies. Maya art often shows kings, queens, high-ranking officials, and priests engaging in ceremonies of auto-sacrifice. They would pierce their tongues or penises with obsidian points

---

* Their bodies had been left on the cave floor, without any grave goods. Some of them had fractured skulls or other injuries showing that they had been deliberately killed.

or stingray spines, then collect their own blood in bowls or on pieces of paper, which were then offered to the gods, sometimes by burning. Awe's interpretation of the stelae is reinforced by the presence of these two bloodletting points, found at their base.

The Stelae Chamber seems to have been a place where such rituals were performed, presumably by very high-ranking individuals (5). Judging by the artifacts they left behind in this chamber, as well as similar ones found in other caves, the Mayas would likely have offered their blood within bowls that they subsequently smashed on altars. They may also have offered sacrifices of animals and incense as part of the ceremonies.

It's impossible to say with complete certainty what the purposes of the rituals in ATM were. I've seen suggestions that they were important in maintaining relations with one or more beings from a complicated pantheon of Maya deities; perhaps they were used to commemorate some important event, to fulfill obligations necessary to maintain humans in harmony with the universe on a particular day of the calendar, or to plead with the gods for rain during a drought. A stone tablet found within the Stelae Chamber engraved with the image of a being with a fanged mouth might indicate the identity of at least one of the intended recipients of these offerings. The fanged image seems to depict a god of rain and thunder, like Chaac or Tlaloc, who were deities deemed crucial for the successful growth of the crop upon which everyone's lives depended, maize. Both the Maya and Aztecs regularly performed sacrifices to this god for the survival of their people. Like Abraham in the book of Genesis, they sometimes offered what they valued the most—children—in accordance with the god's wishes (6).

During the time that ATM was used, the Mayas lived in city-states across a territory that spanned the eastern portion of Mesoamerica and included southern Mexico, Belize, Guatemala,

El Salvador, and the western portions of Honduras. Each of these city-states and their associated territories was ruled by individual royal dynasties, supported by high-ranking nobles and priests. Scholars from each dynasty wrote texts commemorating histories of warfare, accessions, and the deeds of great hero-gods. Astronomers tracked the movement of the planets and stars, keeping careful track of the calendar cycles and helping tie both mundane and spiritual activities to their proper sacred days. Skilled artisans created luxury items and stunning artworks, which, along with food and valuable raw materials like jade, copper, and gold, were traded by merchants via long-distance networks spanning Mesoamerica. Architects designed massive temples and palaces at sites whose names ring out in history: Tulúm, Ek B'alam, Copán, Palenque, Tikal, Chichén Itzá. All of these specialists were aided by mathematicians, who created the most sophisticated numbering system in the world, including the invention of the concept of zero.

Sustaining these elite classes were thousands of commoners in each kingdom: the people who cleared farmland out of rain forests; raised maize, beans, and squash; and provided the labor force for mining and construction.

The ancient Mayas shared many beliefs and cultural traits with other civilizations throughout Mesoamerica, including the sacred ball game,* a calendar system, writing, hierarchical political

---

* Ballcourts are common features throughout ancient Mesoamerican sites. The games that were played at these sites likely differed slightly from one another and the version played by contemporary Indigenous peoples, but probably involved two teams passing a rubber ball back and forth in a manner similar to volleyball except using their hips. The game had ritualistic meanings; in the ancient Maya belief system it was played between the Hero Twins and the Lords of Xibalba (the Underworld) in a struggle between life and death. Beginning around the Classic period, human sacrifice—possibly of the losing team—became associated with the ballgame at some sites.

systems, astronomical and scientific knowledge, large city centers, massive temples and palaces, intensive agriculture, and shamanism. These traits were transmitted and reinforced by extensive interactions between the varied and diverse states: trade, alliances, intermarriage, and warfare. The history of the emergence and development of Mesoamerican cultural traits has been intensively studied by archaeologists and other scholars. While they disagree as to the details of how and why different traits arose, their consensus is that all major Mesoamerican cultural elements were widespread by about 400 BCE, the end of the period archaeologists call the Middle Preclassic.

Many people are familiar with this history, as it's commonly taught in textbooks. What most (non-Native) people don't understand quite as well is that the Maya peoples aren't "vanished," "lost," or "mysterious." They and their many diverse cultures still exist today throughout their ancestral homelands. Over 6 million people today living in Guatemala, Mexico, Belize, Honduras, and El Salvador—doctors, farmers, politicians, archaeologists, musicians, domestic workers—speak one of the many Mayan languages, self-identify as Maya, and maintain a connection to their ancestors and history. Many serve as stewards of their cultural heritage as traditional knowledge holders, archaeologists, historians, park rangers, and tour guides.

This connection among the people and their lands, languages, beliefs, and histories was something that Spanish colonizers tried strenuously to sever. They inflicted all kinds of atrocities upon the Maya—as colonizers from other countries did to other Indigenous peoples throughout the Americas—in service of this goal. The countless and diverse histories of the Indigenous peoples of the Americas converge on shared experiences at the hands of European colonizers: attempted genocide, rape, violated treaties, broken promises, and discrimination that prevails through the present, but also resilience, survival, and a connection to lands and heritage that is passed from generation to generation.

Archaeology and genetics show that the many Native peoples of the Western Hemisphere share something else too. The many threads of their histories converge at a point in the far distant past when the ancestors of the first peoples in the Americas moved from Beringia into new lands in North America. Over just a few thousand years, they explored and adapted to environments across more than 16 million square miles of rocky coasts, deep old-growth forests, high plateaus, endless grasslands, lakeshores, and high arctic tundra. They built mobile camps, small settlements, farming communities, and grand civilizations. They lived as hunter-gatherers, ranging seasonally across vast territories. They lived as farmers, fishers, and livestock herders. They developed sophisticated knowledge of ecology, medicinal plants, and astronomy. They passed down this scientific knowledge to their descendants, along with stories, songs, and languages.

The histories of these peoples are told in the things they left behind: gigantic mounds made of earth, apartment houses perched in caves high above valleys, elaborate stone pyramids, networked roads that linked towns, an isolated hearth out on the high plains, a small sandal left behind in a desolate cave decorated with elaborate rock art, a projectile point embedded in the rib of a mammoth, the skids of an ancient sled. These objects, constructions, and artwork tell us about the countless societies that flourished, diminished, or continued into the present day across the American continents. Each society developed sophisticated ways of living in different environments, from the elaborate irrigation canals created by the Hohokam to water their crops in the Sonoran Desert about 1,100–500 years ago to the *igluvijait* (snow houses) that allowed the Thule to thrive in the long bitter winters above the Arctic Circle a thousand years ago.

Their histories are also told in their genomes. Although contemporary Native Americans are genetically very diverse, with ancestries from all over the globe, their forebears could trace biological ancestry to one (or a very few) founder populations (7). Many contemporary Native Americans have these ancient signatures in their genomes as well. In the last few decades, these genetic legacies from their ancestors have become recognized as a source of information about the past fully as important as artifacts and structures. Researchers—unfortunately, almost all non-Native—have rushed to sequence DNA from contemporary and ancient Native Americans in order to understand the secrets those genes have to tell.

As we discussed in earlier chapters, anthropologists and historians once thought the prehistory of the Americas constituted a single entry event, one that we could use as a starting point to understand the more complex population histories elsewhere in the world. This event started toward the end of the Pleistocene Ice Age, when temperatures were so cold that much of North America was covered by massive ice sheets. Sea levels were so much lower than they are today that Asia was connected to North America by the Bering Land Bridge. The Ice Age ended around 13,000 years ago; during its waning years the Earth warmed enough that the ice sheets began to melt and a thin corridor between them ran down western North America. A small group of people migrated rapidly from Siberia across the Bering Land Bridge and then down through this corridor into the ice-free regions of central North America. They may have been following herds of mammoths or bison, whose bones are sometimes found with finely made 13,000-year-old spear points—called Clovis points—embedded within them. These Clovis peoples were initially few in number, but as they moved throughout the previously unpopulated lands, they increased in numbers and eventually gave rise to all Indigenous peoples in the Western Hemisphere.

In the last 10 to 20 years, however, a mountain of new evidence has emerged, showing us that much of this view of history is wrong. We discussed archaeological evidence in the last chapter. Genetics has also played a major role in this paradigm change, and every year new findings show us that the early human history of the Americas was more complicated than we could have possibly imagined.

Belize provides us with an example of how genetics has complicated simple models of history. The initial peopling of Central America has long been misunderstood. Compared to the wealth of information that later civilizations left behind for historians to ponder, the first people in Central America left a very light footprint, and much of the evidence that they did leave isn't well dated. Many maps showing projected peopling routes have a vague arrow running from North America through Mesoamerica, generally implying that there was a single migration of people southward through Mexico.

But over 9,000 years ago, long before the Mayan civilizations arose, an elderly woman was buried within a rockshelter not far from Actun Tunichil Muknal whose genome tells a different story. She was closely related to an ancient Clovis individual— whom we will talk about more later—over 3,000 miles away in Montana. When researchers compared her genome to those of contemporary Maya, they found that there wasn't a simple ancestor-descendant relationship as might have been expected. Two men buried 2,000 years later in the same rockshelter helped the researchers reconstruct what had happened. Sometime between 9,000 years and 7,400 years ago, a new group of First Peoples moved between North America and Central America. They married into the Mesoamerican communities that were already there, spreading their DNA—and undoubtedly language and culture—widely. The genomes of contemporary Maya reflect descent from this intermarried group of people (8).

This study and another that showed a similar picture from the genomes of other Mesoamericans (9) raise more questions than they answer. Who were these new people (labeled "Unsampled Population A" by researchers)? Where in North America did they originate? What prompted this migration?

We're working hard on answering these questions. In the meantime, using ancient and contemporary DNA, we have pieced together remarkable stories about how people first came to the continents currently called the Americas. In the coming chapters I'm going to be telling some of these stories about their struggles against the odds to survive and how they came to thrive in environments previously unknown to humans.

Today geneticists are also asking new questions: How did evolutionary and cultural forces in the past affect ancient Indigenous societies? How did these events in the past shape the lives of contemporary peoples? How mobile were people in past societies—where did they go, how did they get there, and what did they experience along the way? How did biological and cultural adaptations allow people to survive in extreme environments? How did people deal with the effects of a changing climate and avoid extinction?

These are important questions for geneticists. However, due to the history of colonization and sustained discrimination, many topics in Native American genetics are fraught with political and social implications and tainted by a history of exploitative research. As I tell you genetic stories of the origins of the peoples of the Western Hemisphere throughout the rest of this book, I will also be telling the story of *how* these histories have been learned—and why that matters today.

Chapter 5

*For an Indian, it is not just DNA, it's part of a person, it
is sacred, with deep religious significance. It is part of the
essence of a person* (1).
                            —Hopi geneticist Frank Dukepoo

The elevator ride down to the basement of Fraser Hall was excruciatingly slow, and once the doors opened, I was immediately hit
by the eyewatering odor familiar to all ancient DNA labs: bleach.
The smell intensified as I walked down the windowless, dimly lit
hallway to an unassuming metal door, only made interesting by
its thickness and the sign reading WARNING: KEEP OUT! AUTHORIZED
PERSONNEL ONLY! When I swiped my ID card, a green light flashed,
indicating that I was authorized, and the door unlocked. As I
turned the handle and pushed hard against the metal door, a
fresh wave of cold, bleach-scented air blew my hair back and
wafted out into the empty corridor.

Just inside the door was a white mat. I carefully stepped onto
it, transferring the dirt from my shoes to its sticky surface.

This white mat, which was already slightly spotted with the
impressions of several other sets of footprints—most pointing forward into the chambers beyond, but a few pointing
outward—marks the first step in series of rituals starting from

145

the basement hallway of Fraser Hall to the place where ancient DNA is extracted and cataloged.

None of my own footprints from past visits to the lab were on the sticky mat, whose surface is actually covered with replaceable gummy plastic sheets that are changed when the surface gets too dirty. As one of the two principal investigators (PIs) in our lab group at the University of Kansas, I rarely get to work in the ancient DNA lab anymore; our jobs are to supervise projects and students and bring in funding.

But today was a chance for me to get to do the kind of hands-on lab work that I'd performed for over a decade before accepting a job as an assistant professor in the University of Kansas Department of Anthropology. Getting a chance to set aside my latest grant application and get my hands dirty was a rare treat for me.

Not that my hands *could* actually get dirty. In fact, the process of moving from room to room within the lab involves so many strict procedures that it begins before you even step on the elevator; you have to make sure you are dressed in scrubs (like the kind you see in hospitals) or similarly disposable clothing. This attire certainly contributes to the chill, but the temperature of the lab is also kept intentionally very low.

I was thrilled to get a chance to don my scrubs and work with ancient DNA samples again. After the sticky mat stripped the dirt from the bottom of my shoes, I removed them completely and pulled two small booties—the same kind you wear to walk across damp, newly cleaned carpets—over my socks and put my earbuds in. I adjusted the volume of my podcast higher than would normally be comfortable for me, but I'd learned in the past that it was necessary in order to drown out the sound of air blowing constantly through filters out of the lab. I pulled on a hairnet, making sure that it was fully covering my ears as well as my long hair.

The antechamber where you change your footwear is called the anteroom. We store supplies in this room and use it as a transitional space for putting on booties and hairnets.

As I pulled open the next door, above my head a small red ping-pong ball in a little plastic tube spanning the doorframe rolled from my side of the doorway to the inner side. Air blew into the room I was entering, and it would continue to do so until the door was firmly shut behind me. We did not want air blowing into the lab for very long, and so I stepped onto another sticky mat (this one much cleaner) and closed the door as quickly as I could. The ping-pong ball rolled back, indicating that the airflow had returned to negative pressure and that the air was once again blowing steadily from the innermost room in the lab back out into the corridor. This is characteristic is unique to ancient DNA labs; most clean rooms used for working with pathogens maintain a negative airflow, sucking air into the lab to prevent bacteria or viruses from getting outside. But our laboratory has the opposite concern. We want to prevent the DNA that people are constantly shedding from getting into our lab. The intact and plentiful modern human DNA outside of our clean room would immediately swamp any ability to detect and study the scarce, tiny, fragile fragments of DNA we have coaxed out of ancient bones and soil. Thus our air blows outward into the corridor, the first line of defense against the ever-present threat of human contamination.

As I mentioned, it's cold in the lab, and even colder inside the room I had just entered, known as the garbing room. We keep it cold to prevent everyone working inside from sweating in their protective garb—our second line of defense against contamination. When you're hunched over a laboratory bench moving tiny amounts of liquid from one tube to another with a pipette, even the smallest sweat droplets can pose a serious contamination risk. One drop of DNA from your own sweat could spoil an entire sample, ruining years of work.

Within the garbing room, it was time for me to suit up according to a specific choreography. I put on a pair of gloves before I pulled on a "bunny" suit. A bunny suit is a bit like adult footsie pajamas: a full-length, full-sleeved garment enveloping the entire body, from step-into footed pants to the large hood that is pulled low over your forehead. We try not to touch them with our bare hands, to avoid transferring DNA to their outside surface.

The next step is protecting the ancient DNA from your face and your breath. I unwrapped a surgical mask and placed it over my mouth and nose, using the tie in the back to also keep my hood from slipping over my eyes. I covered my hands and wrists with new gloves, carefully pulling them over my sleeves so there was no gap in coverage over my wrists. Finally, I pulled a sleeve guard—which is essentially a long tube of material with a large opening at one end and a smaller one at the other—over my hands and arms. This provides an extra layer of coverage over the wrists in case my gloves might slip.

By the time I was completely garbed, I looked like a ninja; the only part of me exposed to the air was a thin strip of my face from the bridge of my nose to just above my eyes (some people even go so far as to wear extra-large goggles as an additionally protective measure, but these can fog up from the warm air collecting in your mask and aren't particularly necessary). Although it had been a while since I'd gotten the chance to bunny-suit up, I have been doing this a long time and have learned to trust my equipment and abilities. If my suit did its job, none of the skin cells that I shed would get onto the equipment or bench surfaces.

Just to make sure that I didn't have any DNA on my suit, I spritzed myself all over with one of the spray bottles filled with diluted bleach, closing my eyes against the searing mist. With my eyes still closed, I rubbed my arms, legs, torso, hands, and head with a special towel kept solely for this purpose. Then I waited, listening to the air flow from the overhead vents, counting the seconds until I knew enough time had passed for the air to have diffused the bleach and I could open my eyes again. The smell remained pervasive and intense, but you cannot be an ancient DNA researcher with any degree of bleach sensitivity; the stuff is ubiquitous in our line of work. Bleach—or more specifically, sodium hypochlorite—is a strong oxidizer. It's the third major line of defense against contamination. We spray everything that enters the lab with diluted bleach, and we bleach working surfaces and equipment constantly. The very few visitors we allow into the lab sometimes comment on the "messiness" of our rust-stained equipment and white-tinged benchtops, but this is a sign that the lab is as clean as it should be. Bleach is the reason for coming to the lab in surgical scrubs; I'd learned after many ruined outfits in graduate school that the bleach often soaked through the bunny suit. I'd also learned on lab days that even after I left, bleach was my lingering eau de parfum.

Once I was properly decontaminated and able to see again, I opened a large metal cabinet in the corner of the garbing room. Inside the cabinet was a stack of deep drawers, labeled with names like "Aleutian Islands," "Elk project," "Kanarado soil." And inside of each drawer lies history. Hundreds of samples are in these drawers: of bone, of teeth, or of soil obtained from archaeological excavations.

These samples, all of which have been obtained with the permission of the present-day descendants or stakeholders, have the potential of unlocking the history of how the Americas came to be peopled. Though the process of readying oneself for a day of work in the lab may involve intricate and complex procedures, I have often thought that the effort it takes to achieve a level of cleanliness to open one of these drawers means so much more than just garbing up properly.

The process of requesting permission, removing one's shoes before entering the room, preparing special clothing, misting myself down, and even closing my eyes were all actions that ritualized a specific mindfulness.

This mindfulness is crucial. The human remains in these drawers hold a tremendous amount of significance, and not simply in terms of scientific discovery. These remains represent an acknowledgment to accept responsibility for past transgressions and unscrupulous methodologies, to accept responsibility for preconceived assumptions about race and societies, which resulted in cultural erasures and persisting prejudices. We have promised to treat the small scraps of bone and teeth with respect and mindfulness that they are cherished ancestors, not "specimens," who have been entrusted to us to handle with the reverence they deserve in death. The remains in our lab are the result of a contract between ourselves and the Indigenous peoples who have given us permission to conduct this work, for the early career scientists working now and the scientists they will train in the future, to transform the fields of anthropology and genetics, imbuing them with better ethical practices and a greater respect for human dignity.

As I pulled open a drawer marked with the name of a North American tribe (2), I channeled my sense of amazement at being a part of this work into deliberate focus. Inside the drawer were tidily arranged rows of plastic bags, each marked with a string of letters and numbers. I selected one of the bags and bleached the outside of it, preparing to move it from the garbing room to the DNA extraction room. Since the lab has been constructed for maximum sterilization, each time you enter the next workroom, you are also moving into an increasingly pressurized room, and it is vital that every object you come into contact with is decontaminated. Again, I crossed the threshold, monitoring the ping-pong ball pressure indicator as a sign of the lab's readiness.

Today I was going to try to recover DNA from the tooth of an ancient individual who had been accidentally unearthed as part of a construction project. As was required by law, construction had paused while archaeologists evaluated and eventually excavated the site in order to rescue the other remains and artifacts from the site and to learn what they could before the construction destroyed it. Although the site itself was eventually identified as belonging to a group of people ancestral to a specific tribe, it took some time to determine this, as very little was known about the few scraps of bone that belonged to the people who had been found there. This particular group of people hadn't been buried in graves; instead, their fragmentary remains had been scattered in various places across an assortment of sites: in trash heaps (archaeologists call them middens), underneath the floors of houses, and in the dirt that had gradually covered the sites. After a long time in the custody of a museum while people worked to determine their affiliations, they were repatriated back to the tribe that was descended from them. They were slated to be reburied by the tribe as soon as the proper procedures and ceremonies could be organized.

However, the tribe was interested in learning what they could from their remains. A tribal representative contacted me and asked if it would be possible to use DNA from these ancestors to better understand their history. He had some particular questions about tribal history that couldn't be fully answered with other lines of evidence.

He and I agreed that ancient DNA was an approach that could answer the tribe's questions. Human histories are archived in our DNA, which serves as a faithful, if somewhat cryptic, record of our ancestors' marriages, movements, extinctions, and resilience. By studying the genomes of present-day peoples, we can trace many of the events that affected their ancestors and shaped their own genetic variation. We can use DNA from ancient and contemporary populations to understand Native American population histories in different regions throughout the Americas.

Unfortunately, studying human variation in present-day populations isn't necessarily the best way of examining questions concerning the past. Many other things may have happened—changes in population sizes, intermarriages with other groups, migrations to new places—that could obscure the DNA record of ancestors' histories (3). It is ancient DNA, obtained from bone, teeth, hair, dried tissue, and even soil, that gives us a direct window into the past. The genomes of the ancient individuals from the ancestors of this tribe might help us reconstruct a *genetic* model of their population history. With enough DNA from enough people, we would be able to estimate the size of their population and how it might have fluctuated over time. We could potentially detect gene flow from an outside group into the ancestral population and events that might have happened if marriage and migration accompanied trading relationships. In combination with other types of evidence, we might be able to determine whether one or more individuals from a site came from somewhere else. This information could potentially tell us a great deal about the ancient cultural practices of these ancestors.

Answers to these questions interested the tribal representative, who is an important knowledge holder of their history. They had extensive historical records and oral traditions, but he had some specific questions that genetic data might be able to answer. The tribe seemed to me to be a bit less interested in other questions that I proposed, such as connecting the history of their tribe to that of the Indigenous peoples of the Americas on a larger scale, but they were willing to let me investigate them. After a few years of discussion, we came to an agreement on how the research would be conducted, how the ancestors' remains would be treated, how the resulting information would be shared with the tribe and the scientific community, and how the raw genetic information would be stored to ensure the proper respect for the tribe's sovereignty and privacy.

It had been a long time since I'd worked at the bench, but as I began the process of decontaminating the surface of the 500-year-old tooth—a soak in bleach, a rinse with DNA-free water, a 10-minute session in the small ultraviolet light box on the benchtop—I was relieved to find I still had my "hands." Anti-contamination laboratory practices in the ancient lab are exacting, but I have found that they also translated into excellent preparation for a coronavirus pandemic. You must never touch your face with your hands. You must never pass your hands over open tubes, nor leave tubes or containers with their lids off a moment longer than necessary. Every time you touch your hands to any surface, you must bleach them afterward. Every time you finish working in a laboratory space, you must bleach both the benchtop and the equipment you used. It takes a constant mindfulness and hours of training to operate in this environment. Laboratory researchers call anyone who is skilled at working at the bench a person with "good hands." In the ancient DNA world, this is mostly focused on someone who can maintain this mindfulness in preventing contamination.

This obsessive attention to sterile technique is only one reason why few people want to work in our field. Another reason is how seldom you succeed in actually getting DNA from a bone. While you can obtain huge quantities of DNA from just a swab of a living person's cheek, ancient DNA is a completely different story. Damaged, fragmented, scarce, and mixed with huge quantities of contaminating modern DNA, the molecules are rarely ever present in detectable quantities within any given bone or tooth, and the process of recovering them is extraordinarily difficult.

My confidence in my hands grew as I moved the cleaned, dried tooth to a large cabinet on a nearby bench. This hood, made of clear plexiglass with a hinged front that opened just enough for me to slip my hands into, was effectively an ancient DNA lab-within-a-lab. We had several of them scattered throughout the room, each dedicated to a few stages of the DNA extraction process. Segregating our activities to different enclosed spaces provides yet another critical measure of protection against contamination. But there was even more care that needed to be taken with the first step of the extraction process; mechanically powdering the bone or tooth sample could result in material getting all over the lab. This meant I would need to operate inside an even smaller space within the hood: to be specific, a small plastic glove box that would catch any loose powder. I snaked a drill through one of the ports on the side and began to carefully work on the tooth. The surface material of the tooth went into a small plastic tray, to be discarded as likely contaminated. I widened the hole I was drilling and scraped powder from the inside of the tooth into a second tray (previously decontaminated by exposure to ultraviolet radiation). Taking the tray out of the box, I weighed it on a scale: 0.025 grams—about half the amount of material that I needed. I continued to drill into the tooth, trying hard not to crack it as I excavated powder. This was my least favorite step in the whole process—I was tense and utterly focused until I had tipped the white powder into a DNA-free plastic tube.

I breathed a sigh of relief when at last I had extracted sufficient powder to move forward. We try not to sample more of an individual's remains than is absolutely necessary for obtaining DNA, and there is tremendous pressure not to ruin the tiny samples.

But the rewards are worth the pressure.

It's amazing when you stop to think about it: This tiny extraction from a 500-year-old tooth, this powder, smaller than a pinch of salt, might contain a record of thousands of years of this person's ancestors. I filled the tube with a solution that contained a chemical to sequester the calcium present in the powder and then added a small amount of an enzyme that would chew up all the proteins present in the sample. I added the same chemicals to a second tube, which would serve as my negative control: a test of whether any DNA was introduced at any of the subsequent stages of the extraction process. If, at the end of the four-day process, DNA turned up in my negative control, I would have to assume that the sample was contaminated as well. I'd then have to evaluate every possible source of contamination—the chemicals, the tubes, the equipment, the benchtop, the water, my own techniques—until I identified the problem and fixed it. Ancient DNA research groups are built upon trust: trust between the descendant communities and the PIs, trust between the PIs and their university administrators, trust between the PIs and the students, trust that one's fellow researchers have good hands, trust that the reagents are DNA-free, and trust that every person will report contamination immediately and do whatever it takes to mitigate it.

I placed both tubes onto a rotisserie inside an incubator and programmed the temperature to be hot in order to activate the enzyme. The tubes began to gently turn on the rotisserie, and after checking for leaks, I began bleaching and cleaning the lab. The incubation step takes about 30 minutes, and since it takes me about 15 minutes to enter and exit the lab, there was no point in trying to do anything else in the meantime. These long stretches of waiting in the ancient DNA lab can be incredibly boring, and without anyone else there to talk to (we generally try to limit the number of people working at the same time to one per room), I learned early in graduate school that podcasts and audiobooks are essential. I shadowboxed my way around the lab while I listened, trying not to get my heart rate high enough that I would begin sweating, but still do the work to maintain a different set of hand skills that—like my lab hands—were used all too rarely these days.

The next step after removing the tubes from the incubator was to change the buffer and enzyme inside the tubes. I put them into a benchtop centrifuge to spin the samples at a high rpm. This has the effect of pulling down anything heavy (including the tooth powder) into the bottom of the tube. I carefully removed the liquid using a pipette and added fresh buffer and enzyme. The tubes would now mix overnight on the rotisserie, warmed to the same temperature as the average human body.

I exited the extraction room back into the garbing room and removed my sleeve guards, mask, bunny suit, gloves, and hairnet. Without the extra padding and the intensity of focus, I felt the cold again. I quickly hung up my suit and threw away everything else. As I pulled the door to the outer storage room open, the air blew my hair over my face and I heard the soft clunking of the ping-pong ball moving in its container above my head. After removing my booties, I checked that the door was sealed securely, then flipped a switch that turned on the ceiling-mounted UV lamps. These lamps, which flooded the empty rooms with an eerie purple-blue light, are far more powerful than the lights used in tanning beds. The ultraviolet radiation shreds any stray DNA on the floors, walls, or countertops over the next eight hours, while the sample in the tube mixed overnight, hidden from the UV within a small cabinet.

The next day, I repeated the ritual: careful entry, donning of garb, spraying of bleach. When I pulled my sample tube out of the incubator, I was happy to see that the liquid was a very pale golden brown, while the negative control tube remained clear. This was a good sign; *something* had happened, although I couldn't be sure that it meant any DNA was present. There's no place for cockiness in our field, because the likelihood of DNA being preserved in any sample is extremely low.

However much care we take, however much we hope, we can't know in advance when we extract DNA from an ancient bone whether there will be anything to look at. Early experiments in obtaining ancient DNA (aDNA for short) from human remains revealed that it is rarely preserved, and never preserved fully intact. Researchers are lucky to successfully get DNA out of half the samples they process, and when they do, that DNA is sliced up into tiny fragments, most under 100 bases (the As, Gs, Ts, and Cs that make up the "rungs" of the DNA ladder) long. To give you a sense of how insubstantial such a fragment is, the entire human genome consists of about 3 *billion* bases.

The problem is that the body's cellular machinery, which constantly repairs DNA whenever it is damaged, stops working after an organism dies. Even assuming that the organism's corpse is fairly well protected from the action of scavengers and the environment, the process of decomposition shreds DNA into tiny pieces and inflicts damage on the bases within those pieces. The amount of DNA that remains in an ancient bone depends on many factors: its age, the temperature of the soil holding the remains since the individual died (the colder, the better), and whether the bone was exposed to water, sunlight, or acids in the soil.

You can't know in advance whether there's aDNA preserved in a sample; you can only find this out after a long series of arduous (and expensive!) laboratory processes.

My graduate student was working on bone and tooth samples from other individuals from the same ancient population as my tooth sample, with the hopes that a few might contain enough well-preserved DNA to study. Unfortunately, it was quite likely that none of them would. There's no reliable way to select among the ancient individuals in a population to find out which ones will have the best-preserved DNA. Doing one's PhD research on a project like this is certainly a gamble, and I know some professors who refuse to let their students even take the risk. But ancient DNA holds stories about history that can't be accessed in any other way. The work of careful researchers in laboratories around the world to recover, document, and interpret patterns of genetic variation in ancient peoples has paid off with tremendous insights into human history. My student was eager to be a part of this work, and I reasoned that it was fine as long as she understood the risks involved.

After the tooth sample had mixed overnight, the second day of my work began: an elaborate series of steps to separate the tiny DNA fragments from the protein and other debris present in this golden-brown liquid. We rely on an amazing knowledge of DNA chemistry that has been painstakingly acquired by hundreds of researchers since the early 20th century. Under certain chemical conditions, DNA readily binds to silica, the major component of sand. By running the solution through a silica column, I could hold the fragile molecules in place and clean them with different buffers and alcohols.

Years ago the prevailing methods for DNA extraction demanded that we wash the column of molecules rigorously to remove chemicals in the soil that inhibited later steps. Unfortunately, this approach meant that we generally lost the tiniest chains of the ancient DNA fragments. If I had tried to sample this tooth 10 years ago, it's unlikely that I would have been able to obtain as much DNA as I could now. It has taken some time to improve DNA extraction methods to the point that we can recover more of these crucial puzzle pieces.

One final spin to dry the column, and then I added an elution buffer, changing the pH that was used to detach the DNA fragments from the silica. The finished product looked unimpressive; holding a small tube up to the light, I could see 100 microliters of clear fluid at the bottom, the approximate size of the tip of my three-year-old son's pinky finger.

I carried the tubes into the innermost room of the lab. This was the tiny space past the garbing room and the extraction room kept at the highest positive pressure level of the lab complex. We reserved this room for activities involving the extracted DNA and negative controls. Inside one of three benchtop hoods I began to add tiny quantities of chemicals into small tubes to facilitate the next step in the process: determining whether there was any DNA present in my extraction and negative control. The chemical reaction that I was going to employ in this step—called the polymerase chain reaction, or PCR—would produce millions of copies of a small region of human mitochondrial DNA. In this case, I would target a section of the mitochondrial genome that did not code for any proteins. I would melt the DNA double helix into its two strands with a machine that controlled temperature. Each strand would then serve as a template for making a new strand within a cocktail of chemicals that imitated the DNA copying mechanisms that cells use for reproducing their own genomes. I would target a small region—less than 100 DNA bases long—for amplification with two short, custom-made DNA fragments called primers. The primers would bracket the targeted region and help direct the DNA copying mechanisms to that site (4).

I also added another set of negative controls to this stage so that I might be able to distinguish between contamination that occurred during this step of the process and during the extraction process.

As I squinted at my pipette tips to verify that I was doing things correctly, I remembered the words of James José Bonner, a kindly molecular biologist who took me on as a lab assistant when I was a high school senior. "Most of this job," he told me, "is about moving minuscule quantities of liquid from one tube to another." He was absolutely right (5). Lab workers in the ancient DNA world add additional layers of anticontamination tedium on top of this approach. The people who don't love benchwork often end up specializing in the analysis side of science, like programming and modeling (6). Some impressive souls do both. After my inspection, I carefully closed the lids of the tubes so that no modern DNA could get into them once outside of the clean room, and carried them through the three rooms, repeating the tedious procedure of bleaching surfaces, removing my protective garb, turning on the UV lights. Nothing happens quickly in the ancient lab.

A memorial made of stone marks the final resting place of Shuká Kaa.

Monk's Mound at Cahokia is the largest human-made mound in North America. Its base is comparable in size to the Great Pyramid of Giza. Construction began on the mound approximately 1,100 years ago and it was abandoned after 721 years ago. Although it has been altered by slumping and human activity, its present-day dimensions are approximately 1,037 ft./316.1 m (north-south) by 790 ft./240.8 m (east-west), with a height of about 100 ft./30.5 m. The multi-terraced platform mound supported a large building, which may have been a temple or dwelling for elites.

A Folsom point between bison ribs

A lithics workshop at the Gault site, showing Clovis artifacts—including blades and bifaces in situ. The Clovis components of Gault date to approximately 13,400 to 12,700 years ago.

These Clovis points were excavated from the Gault site.

The Anzick-1 site has been dated to between 12,700 and 12,500 years ago and is currently the only known Clovis-era burial. Both children from the site were reburied nearby in a 2015 ceremony conducted by representatives of the Umatilla, Yakama, Apsaalooke, Yavapai, K'tanaxa, and other tribes.

An archaeologist excavates under water at the Page-Ladson site.

0   5 cm

A bifacially flaked fragment of a knife recovered from the Page-Ladson site dates to 14,550 years ago.

Archaeologist Heather Smith screens excavated sediment near Serpentine Hot Springs in the Bering Land Bridge National Preserve, Alaska. The site has been used by Indigenous peoples for millennia. Excavations at the site by archaeologist Ted Goebel revealed the presence of fluted projectile points dating as early as 12,400 years ago. These points, belonging to the Northern Fluted Point Complex, appear some 500 years after Clovis. They are evidence of a northward movement of Plains peoples, contradicting the idea that Clovis was developed outside of North America and brought southward.

A microblade core in situ from a Denali complex layer (9,000 to 8,500 years ago) at the Dry Creek site near the present day Denali National Park and Preserve, Alaska. Microblades would have been struck off this prepared core. The oldest layers of the Dry Creek site (Nenana complex occupation) have been dated to about 13,500 years ago, making it one of the oldest known sites in Alaska.

Archaeologist Marion Coe works at the Owl Ridge site in interior Alaska, whose Nenana complex occupation dates to 13,100 years ago. Two subsequent Denali complex occupations at the site date to 12,540–11,430 and 11,270–11,200 years ago. This site helps clarify chronologies in the archaeological record of early Beringia.

Footprints at the White Sands Locality 2 site

Geneticist Krystal Tsosie (Diné) teaches DNA extraction methods at the Summer Internship for INdigenous Peoples in Genomics (SING). One of SING's primary aims is to increase Indigenous peoples' participation in scientific research, integrating their own cultural values with genetics.

The next steps in this process would take place in the "modern" lab, a completely separated space. The closer the two labs, the greater the chance that PCR-amplified DNA could get back into the ancient lab, where they pose just as much risk to contaminating the sample as my own DNA. To prevent inadvertently transporting DNA between the two labs, we have strict movement protocols in place; chief among them is that nobody may, for any reason whatsoever, go into the ancient DNA lab the same day after going into the modern lab. If it's absolutely unavoidable, you must go home to shower and change your clothes and shoes before going in the ancient lab. (I've never encountered an emergency so dire that this was necessary.)

It took me a very, very long time to get a faculty job—like with other fields in academia, job opportunities are decreasing each year, and competition for them is extreme. I am grateful for having gotten my dream job—a professor and principal investigator—every day.

And every time I step into the lab, I feel a little thrill of excitement that I actually get to do this work for a living.

No extraordinary precautions to prevent contamination are needed in the modern DNA lab; all research in this space was carried out either on modern human DNA or on ancient DNA that had already been copied millions of times.

I was the only person scheduled to use the lab today. I slipped the tubes into a special machine called a thermocycler and closed the lid. For the next two hours, the machine would cycle the temperature of the tubes, progressing from very hot to separate the DNA strands, to cooler, which allowed the primers to bind to the single stranded DNA, and then gradually becoming warmer again to allow the regions bracketed by the primers to be copied. If there was any DNA present, I should go from those few scraps of broken strands to millions of copies by the late afternoon. It feels like a miracle every time it works.

The PCR reaction serves two purposes: It screens for the presence or absence of DNA in the samples and negative controls, and it also provides the researcher with some preliminary information about this person's maternal ancestors.

Today I was targeting DNA from this ancient person's mitochondrial genome. Mitochondrial DNA (mtDNA) is found within the small structures in living cells that make cellular energy. These structures have their own circular genomes, made up of just 16,569 DNA base pairs (7). Because we inherit our mitochondria only from our mothers (sperm doesn't contribute any mitochondria during fertilization except in very rare cases), mitochondrial DNA lineages are a record of maternal ancestry. That is to say, your mitochondrial sequence is the same as your mother's, and her mother's and her mother's. Not only does it provide this record, but regions of the mitochondrial genome also accumulate DNA changes slightly faster than the nuclear genome, making it a sensitive and accurate witness to recent evolutionary events. The gradual accrual of these changes is a sort of metronomic tick on a millennial timescale.

More ancient DNA studies have been done on mitochondrial DNA than on any other part of our genomes, because there are hundreds to thousands of mitochondria within each of our cells (as compared to just two of each chromosome). Mitochondrial DNA is thus much more likely to be recovered from ancient bones or tissue than nuclear DNA, which only has two copies per cell.

Mitochondrial DNA has taught us a tremendous amount about human history. But unfortunately, its strengths are linked to limitations. Because it's exclusively maternally inherited, this genome provides only a tiny glimpse of an individual's ancestry. You can, for example, test the hypothesis that a group of people buried together are maternally related using ancient mitochondrial DNA. But you won't be able to tell if they share a common paternal lineage, or are more distantly related using that technique alone. Mitochondrial-based population studies may not truly reflect the breadth of history; we have seen many examples of different migration patterns for men and women.

But mitochondrial DNA studies have been extremely helpful in constructing models for population histories in the Americas, particularly in the early days of our field. Many of these models have been confirmed as accurate when scientists have gone back to retest them with other data.

The mitochondrial lineages present within Native Americans prior to European colonization have been well studied. It became clear that there were geographic patterns in the relative frequencies and distribution of these lineages. For example, in the Indigenous peoples of the Aleutian Islands and in the North American Arctic we see just three major groupings of related lineages (called haplogroups by geneticists): A2, D2, and D4.

Similar mitochondrial lineages indicate a common genetic heritage for all these peoples: the Unangax̂ of the Aleutian Islands, the Iñuipiat of the North Slope of Alaska, and the Inuit of Canada and Greenland. (These haplogroups are also shared with the northern peoples of Siberia, suggesting a common origin and/or intermarriage, which is also supported by linguistic similarities.) However, there are differences in the frequencies of each haplogroup in each population; the Unangax̂, for example, have high frequencies of haplogroup D2, and lower frequencies of haplogroup A2. The Iñupiat have a different pattern:

high frequencies of A2, lower frequencies of D4, and almost no occurrences of D2 (I found the first examples of D2 in a study I published with some colleagues back in 2015) (8). These differences reflect different population histories, which is also true for populations all across the Americas in the present and in the past. We frequently see differences in mitochondrial DNA haplogroup composition between ancient and contemporary populations. Some of these differences are because new people migrated into an area, mixing with or displacing the previous inhabitants. Others simply reflect the effects of time.

We can determine whether two populations intermarried using mitochondrial lineages. We can also estimate the opposite: how long it has been since two populations split from each other.

Imagine you belong to a large family with eight siblings. Your youngest sister gets into a fight with the rest of you, and she, her husband, and their kids move to another country. Their kids marry their neighbors and have their own kids, and so forth. Your two families never reconcile and remain in their separate countries. After a hundred years, you would effectively be different families, albeit sharing a common ancestor.

Now imagine the same thing happening (hopefully without the acrimony) to two populations. At some point there was a single group whose members married freely and had children with each other. But at some point in history they split from each other; one portion of the population stopped having children with the other portion of the population. (This often happens because people move away.) Because we can estimate the rate at which DNA bases change into other bases, we can count up the number of randomly accumulated differences between two closely related lineages, and then work backward to figure out how long it has been since they split from each other. That means that, on a macro level, we can determine approximately when two populations separated from each other. This dating approach, using the "molecular clock," isn't perfect by any means, but when interpreted cautiously, it does get us close enough to an approximate date of a population's migration. The molecular clock has had enormous influence on our understanding of human history in general and, as we'll see throughout the book, the prehistory of the Americas in particular.

Two hours later, I returned to my lab and found the thermocycler making a steady humming sound. That was a good sign; it meant the temperature was holding steady at 39°F (4°C), keeping the DNA samples at the same temperature as if they were on ice. The liquid within the tubes looked exactly the same as it had when I put them into the thermocycler; the only way to tell if the reaction had been successful was another procedure, which would take another couple of hours.

Before I had left the lab, I'd prepared for the next step by making an agarose gel, which is essentially a small, clear, slightly squishy square block that closely resembles gelatin. I placed the gel into a special box with electrodes at each end and added enough liquid buffer to just barely cover the gel. At the top of the gel was a series of indentations, or wells. Into each of these wells, I had pipetted a small amount of my PCR reactions along with a tracer dye. To the last well, I also added a ladder, a bit of DNA chopped into segments of different sizes and mixed with dye. Because DNA is negatively charged, when an electrical current is applied across the gel, the DNA fragments migrate toward the positive end of the current. The fragments slowly migrate through the gel in a straight line downward from the well. The smaller the DNA fragment, the faster it will move, and given enough time, all of the fragments in the ladder will separate by size with the largest on top (closest to the well) and the smallest on the bottom. I can then compare the size of these fragments in the ladder with the size of the PCR products that are present in each of my samples. If my PCR reactions worked as they should, there would be a small band present in each well (but not in the negative controls) that would correspond to the length of DNA between my primers.

That was the plan. I double-checked to make sure the gel was oriented with its bottom pointing toward the positive electrode and turned on the power supply. The first sign that things were working correctly was the presence of small bubbles emerging at either end of the gel box. It meant the current was running. I kept watching for a few more minutes until I saw that the dye had moved a smidgen down from the wells—this meant the current was oriented correctly and that the DNA would be migrating in the correct direction—and I turned to my computer to do some other work.

After an hour had elapsed, the gel now displayed two rather pretty bands of color: dark blue at the bottom and a fainter purple in the center. My PCR products ought to be somewhere in between them, and by this point the gel should have run long enough to separate the ladder bands sufficiently. I carefully placed the gel onto a clear glass screen inside of a machine, closed the door, and flipped a few switches on the side of the machine. An image of the gel appeared on the tablet screen in front of me: a dark square with one faint band about three-quarters of the way down from one well, and a ghostly glowing ladder at the far end. I checked—the band belonged to one of my samples and was the correct size. I checked again—nothing lit up in any of my negative controls. Success! My sample had something that amplified with mitochondrial primers, and I hadn't contaminated it. I still had to wait several days for final confirmation from a company to which I'd sent a portion of my sample to sequence the DNA bases, but things were looking good.

Amplification and sequencing of mitochondrial fragments is a bit of an old-school way of studying ancient DNA. As I mentioned before, although mitochondrial DNA can tell us a lot about a population, it does have some limitations. When methods for reading the DNA sequences of complete nuclear genomes—all 3 billion base pairs—became available, the ancient DNA community quickly realized their potential. It took time to optimize the DNA extraction methods for these approaches. Talented researchers identified the skeletal elements that were likeliest to contain enough DNA preserved for whole genome sequencing, which turned out to be teeth and a small pyramid-shaped section of bone at the bottom of the skull that contains the inner ear bones (called the petrous or rocklike portion, in recognition of its density). Other researchers perfected the extraction method that I used, allowing the smallest DNA fragments to be captured and cleaned.

Having ascertained that there was ancient mitochondrial DNA preserved in the tiny drops of liquid at the bottom of my tube, I was back in the lab a few weeks later to make a genomic library. Sitting at the benchtop inside the highest-pressure room, I slipped my arms into the hood and began to carefully mix a small amount of my DNA extraction with water. Keeping all the tubes inside a chilled rack to prevent the reaction from starting before I was ready, I added a few drops of a buffer and a tiny amount of a powerful (and outrageously expensive) solution and pipetted the solution a few times to make sure it was completely mixed. I then placed the tube inside the thermocycler we keep inside the ancient lab and pressed a few buttons. After returning the reagents to the freezer, I settled back in my chair to wait.

For the next 45 minutes or so, as I was absorbed in a podcast, the machine slowly warmed the solution, first to room temperature and then to a blisteringly hot 145°F (63°C). Inside the tube many millions of random DNA fragments that had been extracted from the sample were having their ends "repaired."* Think of each tubeful of DNA as containing millions of tiny little snippets of ribbons that you want to use for an art project, but they have been chewed up by a rodent. Looking at the ribbons under a powerful magnifying glass, you can see that the rodent has gnawed them into a myriad of different sizes. Some have toothmarks on them; some have their ends frayed. In order to do anything creative with them, you'll need to do something about those messy ends. The easiest way to clean them up is to snip off the jagged parts, so that each ribbon has a tidy, vertically straight end.

Although I couldn't see it, I knew that this is what the DNA I extracted from the sample looked like. The ends of each DNA fragment are ragged from the process of degradation that makes ancient DNA so difficult to work with. This part of the process of library creation chemically prepares the ends of each DNA fragment to attach to other DNA fragments, called adaptors.

The next steps I would do in the lab that morning would be to add (or ligate) two kinds of very short DNA sequences to the ends of each DNA fragment. To continue my ribbon analogy, through a series of steps I'd glue an identical little bit of green ribbon to one end, and a little bit of yellow ribbon to the other end of each of those tiny, tidied-up fragments in my tube.

These little bits of ribbon/DNA sequences contained primers

---

* When working with intact DNA from modern sources, you have to actually chop it up into tiny fragments before creating the libraries. We obviously don't need to do that with ancient DNA, which is already extremely fragmented.

that would allow me to amplify and barcode all the library fragments. This would enable me to distinguish between the DNA that I'd extracted in our ultraclean room and anything that had entered the tube as soon as it left the room.

Over the next few hours, I laboriously added more tiny volumes of liquid to the tubes, mixed them, placed them in the thermocycler for incubation at various temperatures, returned the small but outrageously expensive tubes of reagents to the freezer, and prayed that I wouldn't make a mistake or contaminate something. As I did so, I thought a little enviously of the major lab groups who have fancy robots to automate most of the extraction and library preparation steps and armies of staff that allow them to analyze genomes on a scale and at a pace we couldn't even dream of achieving. They employ brilliant researchers who developed the new methodologies for retrieving and analyzing ancient genomics, changing the field forever.

The downside of trying to study the nuclear, as opposed to the mitochondrial, genome is that ancient nuclear DNA is so scarce that it's often impossible to complete the genomic jigsaw puzzle. Either there simply aren't enough molecules preserved, or the ancient human's DNA is so scarce compared to all the other DNA in the extraction (from soil microbes and other sources) that it would be prohibitively expensive to sequence to levels high enough to get all the pieces needed to assemble for a whole genome. Imagine trying to put together an extremely hard 1,000-piece jigsaw puzzle when its pieces are mixed into a pile with 100,000 other pieces from 100 other puzzles, and every piece that you pull out of the pile costs you money. This is a pretty accurate description of what it's like to try to sequence a nuclear genome—often 1% or fewer of the total DNA fragments you get out of an extraction will belong to the ancient human.

You can screen DNA extractions to estimate how much human DNA is present and whether there's enough to make sequencing worthwhile. Screening saves you a lot of money, but the downside is that it means you can only get genomes from a tiny number of individuals with exceptional preservation.

A second approach is to fish for the human DNA out of the pond of all DNA fragments in the extraction, a process called target capture. To do this, you use baits—fragments of human DNA or RNA that bind to their corresponding DNA sequences, which have been preselected by researchers as informative for population differences. These baits are created from a modern human genome, and they are engineered to be attached to biotin (also known as vitamin H), a molecule that likes to attach to a protein known as streptavidin. The biotin-streptavidin binding is so strong that it's used for countless clever applications in molecular biology. In the case of this ancient DNA fishing expedition, researchers coat magnetic beads with streptavidin. When they have baited the pond of DNA fragments, they mix them with the beads. The baits bind to the streptavidin-coated beads, which are then held in place by magnets as they are repeatedly washed. All nonhuman DNA is thus removed, and (after unhooking the DNA "fish" from the baits) the purified ancient human DNA fragments can then be amplified and sequenced. This approach provides data from preselected sites across the whole nuclear genome while being vastly cheaper and more feasible than trying to sequence the whole genome outright. A group of scientists has made the sequence for their capture probes for over 1.2 million spots across the genome (the 1240K capture array) freely available (9).

My libraries were finished. Over the next few days, I verified that the library preparation was successful. Initial sequencing of a small number of the reads from the library showed that the ancient human DNA molecules made up about 10% of the total DNA in the sample, enough to let me sequence the entire genome. As soon as I received the data files from the sequencing run, I gratefully passed them along to a collaborator who specializes in analysis of ancient genomes and tried not to be too impatient as I waited for the results of an analysis of, the first genome to ever be sequenced from this region of the Americas. I worried about many things: that the DNA wouldn't produce a high-quality genome, that one of the larger labs would scoop me (publish a result from this area before me), that I would misinterpret the results, and what the descendant community would think of them. But worrying about things that I could not control got me nowhere, so I tried to push the project from my mind for a few weeks and let my collaborator work in peace.

When the first ancient genomes began to be sequenced, computer programs had to be developed to distinguish between genuine ancient DNA fragments, those of modern contaminants, and the genomes of microorganisms that are co-extracted from any sample. One major breakthrough was the realization that ancient DNA has characteristic damage patterns that distinguish it from contaminating modern DNA.* That meant that by using a particular kind of program, one could estimate the degree of modern DNA contamination and distinguish between damaged and undamaged reads.

Additional methods were developed to match tiny ancient DNA fragments to a map of the complete human genome

---

* Specifically, ancient DNA has a high degree of cytosine deaminations. This causes a particular kind of mistake to be made when the enzyme polymerase copies ancient DNA molecules: on five prime ends of ancient DNA molecules you see a high frequency of Cs changed to Ts, and at the three prime end you see a high frequency of Gs changing to As.

reference sequence. This allowed the book of a person's genome to be assembled from the millions of random sentence fragments that we recovered from their bones. The greater the depth of sequencing—the more times we had a word or sentence confirmed by multiple fragments—the more confident we could be in distinguishing between the true DNA sequence and damage. Computational geneticists invented new ways of comparing ancient genomes with other genomes from ancient and contemporary people and developed powerful tools for modeling population histories using these new kinds of data. Now we could estimate changes in population size at different points in time, detect different sources of ancestry (including from ancient populations that we haven't yet directly observed), and more accurately model past migrations and mixing events (10).

Many of the insights that we have gained into the histories of peoples in the Americas—much of what we will discuss in later chapters of this book—are due to these incredible breakthroughs. The detail about the past that we can now discern is as different from past methods as Google Earth is from a paper map.

Several months after I sent him the data files, my collaborator's email appeared in my inbox. *You have something very interesting here,* the first line read. *This individual belongs to the SNA clade, and closely resembles Anzick-1. I need to do some more work to confirm, but he may have some ancestry from an unsampled population, though definitely not the Ancient Beringians.*

These sentences, which would have been utterly baffling to an archaeologist or geneticist 20 years ago, helped me understand exactly how this ancient person fit within the increasingly complex model of population history that ancient genomes are revealing. In the next chapter, we will begin our exploration of this history.

PART III

Chapter 6

All geneticists and most archaeologists agree that in an evolutionary framework, the question of where the Indigenous peoples of the Americas come from is a biological one, connected to the broader dispersal of anatomically modern *Homo sapiens* (AMHS) out of Africa. This approach places genetic evidence in the forefront of the investigation and then tests the models it produces with archaeological, linguistic, and environmental evidence.

As we discussed in previous chapters, biological data, including physical features, classical genetic markers, and mitochondrial and Y chromosome DNA, show us that Indigenous peoples of America have ancestry from ancient populations in Asia. Geneticists always knew that the models they were working with were oversimplifications of a much more complex process. When we began sequencing whole genomes, particularly from very ancient ancestors, we also began to see just how many details we had missed.

The new story emerging from these new data is infinitely complex and still unfolding. In this chapter, we will explore the genetics piece of this story, noting places where it agrees with and diverges from archaeological evidence.

It's difficult to know where (or when) to begin a genetic history. Genetics complicate a straightforward narrative, and any place you may pick as the "origin" of a population will inevitably be arbitrary. We paleogeneticists often talk about "a people" based on information from a single genome, while also recognizing that this is a ridiculous characterization. Although each genome does tell us about many of a person's ancestors, it can't possibly be a stand-in for all of them. And where is the point at which one ancestral group becomes another?

As you read through the next few chapters—which offer one possible scenario for the origins of Native Americans based on current genetic data—remember that the labels we attach to the populations in all images to follow artificially neaten an extraordinarily complex and tangled family tree.

We are using a dozen or so genomes across 20,000 years of time to attempt to understand the movements, matings, births, and deaths of untold numbers of people. What is quite amazing about paleogenomics is how well it works: Each person's genome is a reflection of thousands of ancestors, allowing us to understand human narrative on a grand scale.

But as you read this genetic chronicle, please do not lose sight of the dignity of the human beings who lived this history and the rich complexity of individual existences that are lost in the telling. The story I tell here is akin to reconstructing a person's entire life by stitching together the photos they posted on Instagram. Not inaccurate, necessarily, just...incomplete.

The images in this chapter present an overview of the history genetics tells us about the first quarter of our story: the formation of the gene pool that gave rise to Native Americans between about 43,000 and 25,000 years ago. Think of them as our roadmap, a way to orient ourselves in the confusing twists and turns of migrations and population formation. Our journey through history in this chapter is aiming for the group or groups that were directly ancestral to Native Americans. To get there, we will have to tell the story of the two populations that served as their antecedents: the ancestral East Asians and the Ancient Northern Siberians.

I want to begin this history about 36,000 years ago with the emergence of the ancestral East Asian population, but to get there, I must first move further back in time. Here is a brief chronological sketch of the events leading up to that point:

Some of populations and population movements during the Upper Paleolithic learned from genetic and archaeological evidence.

Between 50,000 and 34,000 years ago, a period that archaeologists call the Upper Paleolithic, humans who looked like us had left Africa and were rapidly migrating across the globe, from Western Europe to Australia. As they encountered new environments, already inhabited by other kinds of humans we call archaic—Neanderthals and Denisovans—the so-called anatomically modern *Homo sapiens* found ways to adapt to each new place.

They developed remarkable technological solutions to the problems of finding food, shelter, and transportation. They made sophisticated stone tools, including points, blades, and hide scrapers. They also made tools out of bone, including sewing needles, awls, and points. They developed new forms of expressive art: cave paintings, beautifully carved figurines, jewelry. They interacted with their archaic cousins everywhere they went. Some of these interactions led to children, traces of whose ancestry can be found to various degrees in the genomes of all populations of humans today (see the "Genetic Legacies from Archaic Humans" sidebar). About 45,000 years ago, the genome of an individual from the Ust'-Ishim site in western Siberia who belonged to a population ancestral to Europeans and East Asians shows evidence of Neanderthal ancestry dating to perhaps as early as 60,000 years ago.

By the beginning of the Upper Paleolithic, *H. sapiens* populations living outside of Africa were large enough and sufficiently geographically dispersed to accumulate genetic variation that distinguished populations in different regions of the world. The genome of an ancient man whose remains were found in Tianyuan Cave in northern China showed that by about 43,000 to 40,000 years ago, you could tell the difference genetically between people living in western Eurasia and people living in East Asia, although the differences were subtle and there was still considerable mixing between them* (1).

---

* Genetic distinctiveness should not be mistaken for genetic "purity" in any sense.

GENETIC LEGACIES FROM ARCHAIC HUMANS

Anatomically modern *H. sapiens* (AMHS), or humans who look like us, interbred with Neanderthals sometime around 65,000 and 50,000 years ago and with Denisovans sometime around 55,000 and 45,000 years ago (and possibly at other times as well). These introgression events, as geneticists rather primly call them, took place in different regions of the world. Subsequent movement and mixing of people have led to these traces of ancestry dispersing throughout the globe. Currently, traces of Denisovan ancestry are found mostly in populations living in Asia, South Asia, Southeast Asia, and Melanesia. Traces of Neanderthal ancestry are seen at highest levels in East Asians, Europeans, and North Africans. There are low levels of Neanderthal ancestry in East African populations, possibly because of more recent gene flow from AMHS carrying these alleles.

Traces of gene flow from early AMHS have been found in the genomes of Altai Neanderthals. So far, no Neanderthal mitochondrial lineages have been found in AMHS populations, which strongly suggests that these mating events were between Neanderthal males and AMHS females. There's significant evidence that our genomes have been slowly purging themselves of most archaic-derived ancestry generation by generation, but there are a few exceptions. Some alleles that AMHS populations inherited from other kinds of humans appear to be quite beneficial and are positively selected for by evolution to persist in certain human populations.

In 2014, researchers attempting to identify potential

genetic mechanisms underlying the high prevalence of type 2 diabetes (T2D) in Native American populations of Mexico and South America conducted a large genome-wide association study looking for risk factors in the genome associated with the disease. They found two alleles of a gene associated with hepatic lipid metabolism that were high-risk factors for T2D for people from these populations: SLC16A11 and SLC16A13. Researchers identified the source of these alleles in the Neanderthal genome. Their high frequency in these populations might mean that they were beneficial for ancestors confronted with a changing diet. Perhaps the allele conferred an advantage in eating the meat-intensive diet characteristic of living at high latitudes or very cold climates, as the ancestors of Native Americans did for many generations.

Denisovans also contributed helpful alleles to Native Americans. Researchers scanning for evidence of alleles under intense positive selection within the genomes of Native Americans found a stretch of DNA on chromosome 1 that had a suite of variants that were present at high frequencies within Native American populations. This region contained two genes, called TBX15 and WARS2, that are linked with numerous physical traits, including height, body fat distribution, and hair pigment. This allele, which came directly from Denisovans, may have played a role in modern human adaptation to life above the Arctic Circle, perhaps influencing the development of the body types that are common to peoples who currently live in that region (2).

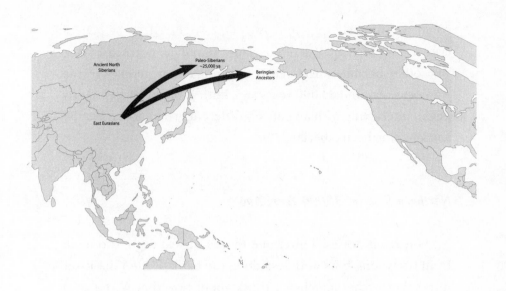

East Asia, about 36,000 Years Ago

Around 36,000 years ago, a small group of people living in East Asia began to break off from the larger ancestral East Asian population. We don't know why or how, but this group began to have fewer and fewer children with their neighbors as time passed. Often this genetic signature signals a population movement of some kind; geographic distance often results in decreased gene flow between the two populations. We don't yet have a good idea of this process archaeologically, but we can see its signature in the genomes of their descendants: a divergence of an East Asian group from the larger population with some reduced gene flow continuing between them for another 10,000 years or so.

By about 25,000 years ago, gene flow with the broader East Asian population stopped completely. Again, we're not sure what happened, but another migration is a possibility. The smaller East

Asian group itself split into two. One, referred to by geneticists as the ancient Paleo-Siberians, stayed in northeast Asia. The other became ancestral to the Indigenous peoples of the Americas. Around 24,000 years ago, both groups independently began interacting with an entirely different group of people: the Ancient Northern Siberians (3).

Northern Siberia, 31,000 Years Ago

All bones tell stories. Thirty-one thousand years ago, two teeth from two young boys were lost along the bank of the Yana River in northern Siberia. When I think about how they were lost, I fancifully imagine a sunny summer afternoon.

The boys were playing in the shallows of the river, shrieking as they scrambled to find which rocks would splash the other more, until one of the adults trying to fish nearby had enough. "Either stop splashing or go play somewhere else," she scolded. "You're scaring the fish, and if I don't catch enough, we won't have enough to eat tonight. You're 10 years old, old enough to nearly be men. Start acting like it!" Sheepishly, the boys retreated to the other side of the riverbank and flopped onto the grass. As they tried to find another game, one of them started wiggling his canine tooth absently—it had been loose for a while, and he loved fiddling with it. Suddenly it popped out! Surprised, he touched the tender spot where it had been, feeling the tip of another tooth poking through his gums. He looked over at his best friend with a sly bloody smile, then tossed the tooth into his lap. "Groooooossss!" his friend shrieked. Pretending to vomit, he reached into his mouth and with a sharp tug, yanked out one of his loose teeth and threw it back in retaliation. The disgusting thing hit his friend's forehead, bouncing off into the grass. Helpless with laughter, the boys pounced on each other, wrestling until they tumbled down the bank back into the river. The teeth, forgotten, eventually sank deep into the mud.

The molar and canine, both worn from use and covered in plaque, were found by Russian archaeologist Vladimir Pitulko 31,000 years later. They show us that these two boys survived the dangers of infancy and early childhood, living until at least age 10 or 11. They were luckier than many; in a time when vaccinations and antibiotics were unimaginable, many children did not live that long.

We don't know the names of the boys or how long they lived. We don't know if they actually played together or fought each other; we don't know if they grew up, married, and had children of their own, or if they eventually succumbed to disease or starvation during the long Siberian winters.

Neither the boys nor their kin left any other bones behind. But we know about them in one of the most intimate ways imaginable; we have their complete genomes, sequenced from the teeth they lost as they grew into adults. These genomes tell us a story of the distant past, of a remarkable people living above 70° north latitude during the early Upper Paleolithic.

The Yana boys' genomes tell us that sometime around 39,000 years ago, their population, which we call the Ancient Northern Siberians, separated from ancestral East Asians. This separation likely occurred as they moved into northeastern Siberia, pushing far past the boundaries of where their distant cousins—the Neanderthals and the Denisovans—lived (4).

From a biological perspective, it's amazing that Ancient Northern Siberians were able to thrive above the Arctic Circle when Neanderthals did not. Neanderthals evolved physical adaptations to cold environments: Their stocky bodies and short limbs were well suited for conserving heat, and the size and shape of

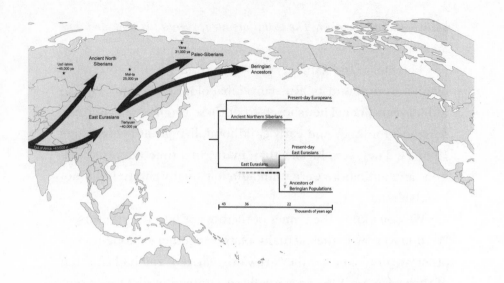

their noses were perfect for warming cold air before it reached their lungs. In contrast, AMHS populations were still adapted to equatorial climates at this point in history, with long limbs and thin bodies, ideal for dispersing heat rather than conserving it.

But AMHS developed more mechanically complex technologies than Neanderthals. Snares and traps allowed them to hunt smaller prey, like birds and hares, which could meet their smaller caloric needs. One piece of technology in particular made the difference between living above and below the Arctic Circle: The humble sewing needle, which today we can buy in packs of 16 for about $4 at a craft store, was a wondrous device 38,000 years ago. The eye in the needle, which required complex planning and dexterous craftsmanship to make from mammoth ivory, allowed people to tailor insulated clothing, sleeping bags, gloves, and house coverings. Imagine standing on the windswept plains of northern Oklahoma on a December evening, or walking by Lake Michigan in the depths of a Chicago winter. Now imagine the temperature about twice as cold. It's easy to see how tailored fur clothing would have allowed you to spend many hours each day outdoors hunting and gathering food, making the difference between life and death in these climates.

Despite the bitterly cold, long winters with average temperatures falling to around −36°F (−38°C), a group of Ancient North Siberians thrived in this region for almost 200 years. They lived in permanent settlements up and down the banks of the Yana River, which archaeologists collectively named the Yana Rhinoceros Horn Sites (Yana RHS).

Thanks to the work of archaeologist Vladimir Pitulko and colleagues who have been excavating at Yana RHS since the 1990s, we know a great deal about the lives of their residents.

They made clothing from rabbit and hare fur. They flaked stone tools and used them to hunt rhinoceros, horses, mammoths, wolves, reindeer, brown bears, and even lions. They carved special vessels from ivory. They wore elaborate necklaces made of ivory beads, some painted with red ochre, and pendants in the shape of horses or mammoths made from amber, teeth, or the dark silver-gray mineral anthraxolite. They made bracelets and hair diadems out of the same materials, and carved mammoth tusks with images of hunters or dancers (5).

Yana RHS was not a small community. The genomes that the team of geneticists, led by Eske Willerslev, sequenced from the two boys' teeth tell us that they were not brothers or cousins, as we might expect from finding the remains of two contemporaneous children from a small population. On the contrary, we can tell from their genomes that the effective population size (the number of breeding adults) was around 500. The actual population size would have been much bigger—perhaps 1,000 people, or more. We're not sure why no burials have been found at the site; perhaps they cremated their dead, or perhaps their cemeteries are still to be discovered.

Eastern Siberia, 25,000 Years Ago

As I read about excavations at the Mal'ta site, in eastern Siberia, this is how I imagine the day the children were buried there.

Only someone who has lost a child can understand how life-shattering it can be. On this day, two families would have laid children to rest in the same grave, saying goodbye to a baby and a toddler. They buried the infant close to the three-year-old. I imagine the mother reaching out a shaking hand to smooth the baby's fuzzy dark hair one last time, before sprinkling them both with the red powder that would help them transition into the next life. The children were well provisioned for their journey, with an ivory point and some flint tools. The toddler's mother had lovingly placed jewelry on him: a bracelet, a diadem, and a beaded necklace with a pendant in the shape of a bird. Perhaps they reflected his desire to fly, lurching after the ravens, chasing them, and flapping his chubby

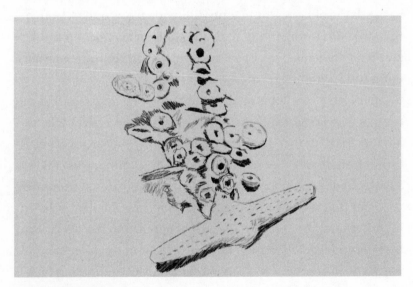

Toddler's Bird Necklace

arms. Perhaps his parents consoled each other with the certain knowledge that his spirit was flying free. After the men covered the sad little hole with a stone to protect it from scavengers, the families trudged back to their camp, forever bound to each other in unimaginable grief.

The children—referred to by archaeologists as MA-1 (the toddler) and MA-2 (the baby)—lived at a site known known as Mal'ta in eastern Siberia near Lake Baikal. Russian archaeologist Mikhail Gerasimov excavated the site in the 1920s, finding an encampment of subterranean houses with bone walls and roofs made from reindeer antlers, which at the time would have been covered by animal skins. Archaeologists have found an astonishing variety of objects at the Mal'ta site, including stone tools, bone projectile points, needles, awls, jewelry, and artworks distinctive to Upper Paleolithic peoples across Eurasia, including so-called Venus figurines—the statue representations of women with exaggerated breasts, legs, and buttocks. The diversity of objects suggests that the people living at Mal'ta were either staying there for extended periods of time or revisiting it regularly; if it were a short-term logistical place like a hunting camp or a workshop, one would expect to see a much more limited range of objects (6).

"The people camped out on the site were the last of the group in the region before the Last Glacial Maximum, the Ice Age," Kelly Graf, a professor at Texas A&M and faculty member in the Center for the Study of First Americans, told me. "They would have been noticing the climate getting colder and dryer."

Graf has been fascinated by the puzzle of where the peoples of Siberia moved to during the LGM and particularly where the ancestors of Native Americans may have come from. The archaeological record of Siberia during the LGM just...stops. There's

A Venus figurine from Mal'ta. Like other Siberian Venus figurines, this one shows evidence for a hood and other warm clothing (Venus figurines in Europe are usually naked).

no evidence of people anywhere from about 20,000 years ago to about 15,000 years ago, when the climate began to warm again. It looks like people throughout Siberia migrated to warmer regions. Or they all died.

Graf and other researchers at the Center for the Study of First Americans had been working on a project with geneticist Eske Willerslev to sequence the genome of the oldest known remains in North America—which we will talk about in the next chapter—when Graf brought up the possibility of trying to get DNA from the Mal'ta children. Having worked extensively in Siberia, she told him about the pre-LGM site and arranged for him to get access to the remains at the Hermitage State Museum in St. Petersburg.

In 2014, a team assembled by Graf and Willerslev and led by Maanasa Raghavan published an analysis of MA-1's complete genome. We know from later comparisons that the Mal'ta boys' people were direct descendants of the Ancient North Siberians from Yana (7). They were broadly ancestral to present-day western Eurasians. But in comparing his genome to those from populations across the world, they found that he was also closely related to present-day Native Americans; his population was directly ancestral to them.* Mal'ta's population—the Ancient Northern Siberians—seems to have encountered the daughter East Asian population described at the beginning of this chapter around 25,000 years ago and interbred with them. Current estimates suggest that approximately 63 percent of the First Peoples' ancestry comes from the East Asian group and the rest from the Ancient North Siberians. We're not sure where this interaction took place. Some archaeologists believe that it occurred in East Asia, suggesting that this is where the Siberians moved during the LGM.

There's also a case to be made for this interaction having taken place near the Lake Baikal region in Siberia from genetic evidence, too. The ancient Paleo-Siberians, as mentioned earlier in this chapter, split from the East Asian ancestors of Native

---

* Thus, in a paradox that is fairly commonplace in paleogenomics, the Mal'ta boy, who did not have any children of his own, is now a genetic representative of the ancestors of millions of people: those who live in Europe, Western and Central Asia, and the Americas.

Americans by about 25,000 years ago. They are known to us from the genomes of an Upper Paleolithic person from the Lake Baikal region known as UKY and a person from Northeastern Siberia dating to about 9,800 years ago known as Kolyma1. Closely related to Native Americans, these "cousin" genomes also show a mixture of ancestry from Ancient North Siberian and East Asian populations, although the proportion of East Asian ancestry is a bit higher than in Native Americans—about 75 percent. The Ancient North Siberian gene flow into the East Asian ancestors of the ancient Paleo-Siberians probably occurred at the same time as into the ancestors of Native Americans—between about 25,000 and 20,000 years ago. Because UKY lived in the Lake Baikal region some 14,000 years ago, some researchers argue, it seems likely that the meeting between East Asians and Ancient North Siberians occurred in the Trans-Baikal region (8).

But other archaeologists and geneticists argue that the meeting of the two grandparent populations of Native Americans—the East Asians and the Ancient Northern Siberians—occurred because people moved *north*, not south, in response to the LGM. (In this scenario, Paleo-Siberian descendants, like UKY, could have been the result of a southward repopulation of Siberia out of Beringia.)

The reason for that is because both mitochondrial and nuclear genomes of Native Americans show us that they had been isolated from all other populations for a prolonged period of time, during which they developed the genetic traits found only in Native American populations. This finding, initially based on classical genetic markers and mitochondrial evidence, came to be known as the Beringian Incubation, the Beringian Pause, or the Beringian Standstill hypothesis (9).

Most geneticists are skeptical that the peoples ancestral to

Native Americans could have been completely isolated for any length of time if they were living anywhere near other populations during the LGM. Thus, we look north, rather than east, for the location of the refugia that may have allowed the ancestors of Native Americans—whom I will call the Beringians for the rest of this book—to survive the Ice Age.

Except for a few small islands, central Beringia is mainly underwater today; it was a substantial land connection between 50,000 and 11,000 years ago. Scientists are drilling into the sediments across this region in order to take cores, whose layers— which contain pollen, plant fossils, and insect remains—provide snapshots of both the geology and the environment across time at each site they are taken. Paleoclimatologists stitch these snapshots together to reconstruct the LGM climate across Beringia, including the regions that are underwater today.

Layers dating to the LGM show us that the environment was patchy across Beringia. There were vast regions of steppe-tundra: dry and cold grasslands, sprinkled with herbs and small shrubs like dwarf willows. This environment, lacking firewood, would have been difficult for humans, although the presence of megafauna across steppe-tundra would provide not only food but also dung and bones to burn. But there were some places in and near Beringia that could have served as more attractive refugia for both humans and animals.

One possible refuge for humans during the bitterly cold Ice Age was the southern portion of central Beringia—a region that is presently under about 164 feet of ocean but would have been lowland coastline 50,000 to 11,000 years ago. Unlike the steppe-tundra regions, the southern coast of the land bridge would have been much warmer and wetter because of its proximity to the ocean.

Paleoenvironmental evidence shows that it actually contained wetlands, with peat bogs and trees like spruce, birch, and adler that people could have burned for fuel. Waterfowl would have visited this place, and they and other animals would have provided a reliable supply of food for the Beringians. This model for Native American origins explains the genetic evidence of isolation. To some archaeologists, it also meshes well with the archaeological evidence. Beringians living on the south coast of the land bridge had access to Pacific marine resources, including kelp, shellfish, fish, and marine mammals. A prolonged stay in a coastal region would have required the population to develop adaptations for these new resources. If true, this period of isolation meant that the First Peoples already had the culture and knowledge needed for thriving in coastal environments by the time routes into the Americas became accessible a few thousand years later. It means that Beringia should more properly be viewed as a lost continent than as a land bridge. The term *land bridge* gives the impression that people raced across a narrow isthmus to reach Alaska. The oceanographic data clearly show that during the LGM, the land bridge was twice the size of Texas. If the Out of Beringia model is correct, Beringia wasn't a crossing point, but a homeland, a place where people lived for many generations, sheltering from an inhospitable climate and slowly evolving the genetic variation unique to their Native American descendants (10).

Recently, environmental reconstructions have pinpointed another region that could have served as a refuge. The arctic zone of Beringia and adjacent areas (the Taimyr Peninsula and the northern portion of Lena Basin), called the Northwest Beringian Plain, were home to a variety of large mammals during the LGM. These animals, including mammoths, horses, saiga, woolly rhinoceros, and musk ox, were extremely well adapted for

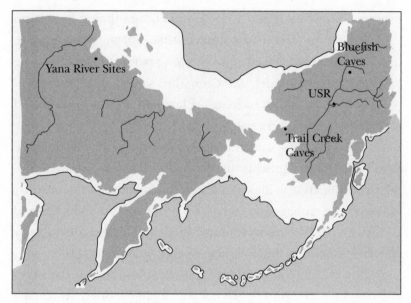

Map of Beringia, showing the extent of land during the LGM and the location of sites discussed here (Adapted from Hoffecker et al. 2020)

the dry steppe-tundra environment of the High Arctic. A large population of predators, including wolves, lived in this environment, and it would also have been a place where large-game subsistence hunters could have thrived (11).

Many geneticists (myself included) argue that evidence of a human presence in south-central Beringia—or perhaps in the Northwest Beringian Plain—during the LGM is there to be found, if we can just look in the right places. There are some tantalizing hints that people were in eastern Beringia well before 14,000 years ago. In lake sediments dated to around 32,000 years ago from the Brooks Range in Alaska, a team of researchers recovered organic molecules known as stanols that looked exactly like those in human poop (12).

The second hint came from deep within Bluefish Cave in the Canadian Arctic. In the 1970s French-Canadian archaeologist

Jacques Cinq-Mars found a few animal bones dating to around 24,000 years ago that had cutmarks that indicated the animals may have been butchered. This was long before the overturning of the Clovis First model, and Cinq-Mars's findings were not well received. His fellow archaeologists laughed at him and refused to take his findings seriously.

In 2015, graduate student Lauriane Bourgeon reanalyzed more than 5,000 bone fragments that Cinq-Mars had put forward as evidence of human activities. While she concluded that the majority of them were damaged from natural processes, she found several bones with parallel grooves cut into them that looked convincingly like those caused by humans, not scavenging animals (13).

Many archaeologists are still quite cautious about the poop and the cutmarks; they're just not convincing enough. And apart from these findings, arguing that there's simply no other archaeological evidence supporting either the Out of Beringia or the Out of the Arctic hypothesis. Nearly all of the locations that could have served as refugia for people are currently underwater at the bottom of the Bering-Chukchi Sea, and the Arctic Ocean north of present-day Siberia. Future underwater archaeologists will someday be able to investigate these areas using submarines, but for now it remains an unanswerable question.

24,000 Years Ago, Somewhere in Beringia?

Either just before or shortly after the start of their period of isolation, the Beringians split into several groups: the Ancestral Native Americans, who would move below the ice sheets and become ancestors of the First Peoples; the Ancient Beringians, who would stay behind in Beringia; and a mystery group (Unsampled

Population A) known to us only indirectly from the traces of ancestry it contributed to some Mesoamerican populations (14).

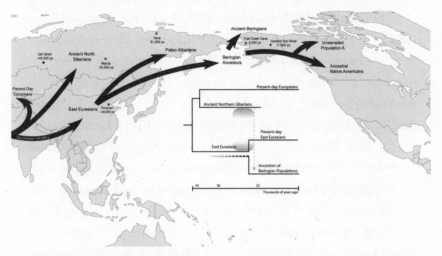

Hypothesized population migrations. Inset shows a phylogenetic tree and time scale, with dotted arrow denoting gene flow.

We have recovered the genomes of the Ancient Beringians from an archaeological site that tells us a story of another heartbreaking loss of children. A little over 11,000 years ago, three children—a prenatal girl, a three-month-old girl, and a three-year-old boy—were laid to rest under a hearth inside their home in the Tanana River Valley in Alaska. Like the Mal'ta boy, they were buried with care and provisions: hunting equipment (including stone points and bifaces) and carved rods made out of antler. The boy was cremated. Like the Mal'ta children, they were sprinkled with red ochre.

We don't know their names, but the peoples who live in the region today—the Tanana Athabaskans of the Healy Lake tribe—call one of the girls Xach'itee'aanenh T'eede Gaay (Sunrise Child-Girl), the other Yełkaanenh T'eede Gaay (Dawn Twilight Child-Girl), and the boy Xaasaa Cheege Ts'eniin (Upward

Sun River Mouth Child). Their remains were discovered by archaeologist Ben Potter in 2013, at a site known today as the Upward Sun River, in the Tanana River Valley (15).

The Healy Lake Traditional Council and the Tanana Chiefs Conference (the broader regional consortium of tribal leaders) were interested in knowing more about them and so gave permission for archaeological and genetic research. "I would like to learn everything we can about this individual (Xaasaa Cheege Ts'eniin)," the Healy Lake Traditional Council's First Chief Joann Polston said in 2011. The Tanana Chiefs Conference President Jerry Isaac agreed: "This find is especially important to us since it is in our area, but the discovery is so rare that it is of interest for all humanity." (16)

The mitochondrial genomes from both the girls and the complete nuclear genome from Xach'itee'aanenh T'eede Gaay showed that they belonged to a population that didn't have any direct contemporary descendants. Their ancestors had split off from the other Beringians sometime between 22,000 and 18,000 years ago and remained in eastern Beringia (Alaska) after the ice wall melted and the other groups moved southward. We can tell from another genome recovered from the tooth of an 18-month-old child found at the Trail Creek Cave site on the Seward Peninsula in Alaska (450 miles away) that this population wasn't small, but rather large and geographically dispersed (17).

We don't know what happened to the Ancient Beringians. The Trail Creek Cave tooth dates to about 9,000 years ago, but present-day people living close to each site are not direct genetic descendants of Ancient Beringians. At some point, a population turnover must have happened, but we do not know the details of that event.

We have indirect evidence that there was another group present in Beringia as well. As we discussed back in chapter 4, the genomes of the present-day Mixe from Central America

show that among their ancestors was a group that was genetically distinctive from the Ancient Beringians and the Ancestral
Native Americans that became the main source of ancestry for
the First Peoples. This mystery group, labeled Unsampled Population A by geneticists, split from the other Beringian groups

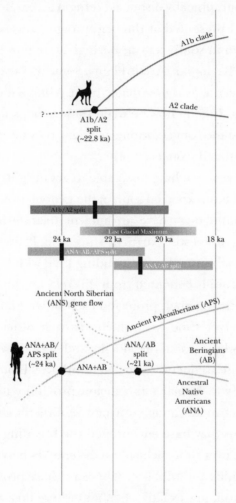

Population splits of dogs (inferred from mitochondria) and humans
(inferred from whole nuclear genomes) during the Upper Paleolithic, with a
timescale for context. Image redrawn from Perri et al., 2020.

sometime before 22,000 years ago. Although a genome from a member of this group has not yet been found, they have left foot-prints in the genomes of their descendants (18).

One possible interpretation for this pattern of rapid splits as the Beringians became isolated is that different groups of people occupied geographically dispersed refugia across Beringia (as in the scenario I presented at the beginning of this section). One may have been in south-central Beringia; one may have been in the northern Beringian/Arctic plains. Some may even have been in North America, as the evidence from White Sands Locality 2 might suggest. If gene flow between these groups was restricted because of distance or geographic barriers, they would have grad-ually differentiated from each other.

In recent years we have been able to get insights into human history from an unexpected source: the genomes of dogs.

Humans and dogs have an ancient relationship; we have co-evolved as domestic partners since the Pleistocene, when wolves presumably first began spending time with humans. Both humans and canids benefited from this relationship: We hunted more efficiently together, protected each other, and found com-panionship. Over time we influenced each other's evolution, with wolves who were more docile and less wary of humans becoming more frequently associated with the hunter-gatherers. The proximity to humans also meant proximity to their food, no small selective advantage during the difficult climate of the LGM. Humans may have encouraged the breeding of the more docile wolves; over time, behavioral differences between human-associated canids and wild wolves became more pronounced.

It's hard to know exactly when and where dogs first became domesticated, as morphological differences between dogs and wolves can be subtle. However, genetics gives us some insights into this process. Dogs were domesticated from a now-extinct species of wolf, likely in Eurasia. Although on separate evolutionary trajectories from wolves, dogs did mate with them fairly frequently throughout history. Accounting for these introgression events, geneticists see dogs as clustering into between three and five major lineages by 11,000 years ago, including a western Eurasian lineage, an East Asian lineage (dingos), and a lineage consisting of ancient American dogs and present-day Arctic dogs and their ancestors.

This latter group is of particular interest to the story of the peopling of the Americas. Genetic evidence shows that American dogs did not evolve independently from American wolves. Instead they show a clear signal of descent from Siberian dogs, suggesting that they were brought to the Americas by the First Peoples. Indeed, the remains of dogs can be seen in archaeological contexts within the Americas at least by 10,000 years ago. If dogs were brought to the Americas by the First Peoples, then perhaps an understanding of the dynamics of dog population history can give us fresh insights into the earliest histories of humans on the continents.

A recent study by Angela Perri and her colleagues took this approach. They compared the branching patterns of dog mitochondrial lineages across the Americas to the currently best-supported models for human population history. Remarkably, they found a close correspondence between the human population history reconstructed from whole genomes (described in this chapter) and the maternal history of dogs in the Americas.

Dogs in Eurasia dating to about 22,800 years ago appeared to have split into two major groups. One group, termed A1b by geneticists, remained in Siberia. The other group, termed A2 clade by geneticists, was ancestral to two daughter groups approximately 16,400 years ago: the A2a clade, which remained in Northeast Asia, and the A2b clade, which moved into North America and gave rise to all ancient dogs in the Americas south of the Arctic.

Compare this history to that of humans, as shown in the figure at the beginning of this section. We see a major branching of the ancestral East Asian population at nearly the same time (24,000 years ago) as dogs (22,800 years ago) into a lineage that stayed in Siberia (Ancient Paleo-Siberians) and another that produced the Ancient Beringians and the Ancestral Native Americans.

Perri and her colleagues have suggested that the Beringian Standstill provided the ideal conditions needed for the domestication of dogs: an environment that favored close associations with humans for access to scarce food resources, isolation from other canid groups, and clear evidence of genetic adaptations for Arctic conditions in parallel with human adaptations. This intriguing hypothesis matches the timing for the emergence of domesticated dogs somewhere between 40,000 and 15,000 years ago (an admittedly wide range of time). Further testing with nuclear genomes should provide more precise estimates of this population history (19).

The story of dogs—and humans—in the Americas does not end in this chapter.

At the end of the LGM, the ice sheets began to melt, and travel southward from Beringia became possible, setting the stage for one of the most astonishing feats in human history: the peopling of North, Central, and South America. In the next chapter we will examine clues from ancient genomes and the archaeological record in an effort to solve the mystery of how they accomplished this.

Chapter 7

Imagine a nondescript corner of what is now Florida, almost 15,000 years ago:

Mastodons used to gather at the edge of the inland pond, far from the coast. The surrounding savannah was dry and hot, but the pond was kept cool by the huge cypress trees and 15-foot limestone cliffs encircling it. This pond was an oasis: a perfect place for drinking water and playing with other members of the herd. The fertile soil around the pond made the ground and trees thick with food: delicious grapes and gourds. The mastodons contributed to this ecosystem, gifting the soil with enormous piles of dung. Imagine if you can—a lush, green paradise, humid with the smoky green scent of cypress trees, rich piles of fresh dung, swarms of buzzing insects, and the earthy smell of wet mastodons rolling in cool water.

The giant long-tusked creatures were not the only inhabitants of this oasis. When a middle-aged mastodon died on the edge of the pond one late summer's day, humans were swift to claim his meat for their hungry families. Using sharp knives made of flaked stone, they butchered his carcass, removing the hide and meat. It would have been messy, slippery work, and it had to be done quickly. The aroma of a freshly dead mastodon would soon attract the big cats and scavengers living nearby, and nobody wanted to tangle with them.

One woman, hacking off huge slabs of meat from the shoulder, snapped the tip of her knife off on a bone. Cursing, she tossed it into the water and pulled out a spare from her belt. There wasn't any point in stopping to reshape the knife—it could be easily replaced from the rock sources nearby, and there was no time to lose.

Mastodon tusks were valuable—full of rich fatty marrow and made of ivory that could be used for tools and jewelry. One of the butchers smashed a hole into the skull just below the mastodon's eyes until he could see the ligament that attached to the tusk. He cut the ligament then held the head steady as another butcher twisted the tusk out of its cavity. They repeated the process for the other tusk.

But the travois was full, and everyone's packs and arms were loaded with meat and hide. There wasn't any room left for the tusks. The butcher couldn't carry both tusks himself, and in his haste to leave the site, he buried one as deep as he could in the muddy bank of the pond. Perhaps the scavengers would miss it, and he could come back for it later. Hoisting the other huge tusk in his arms, he followed the swiftly departing party. They could afford to leave one tusk behind; it would be a good day for their families if they could return home safely with all of this food. There was no reason to press their luck any further.

Scavengers quickly descended on the carcass after the humans left. They removed the rest of the edible parts of the mastodon, leaving bones and the remaining tusk at the edge of the pond. Time passed, and as animals continued to visit the pond, their dung eventually covered the forgotten mastodon's remains. Eventually the climate changed, and the mastodons stopped visiting the area. Soon they vanished altogether.

More than 14,500 years later, archaeologist Jessi Halligan found the discarded broken knife as she sifted through sinkhole layers 30 feet below the surface of the Aucilla River. Halligan and her PhD mentor, Michael Waters, were reinvestigating the site after archaeologist Jim Dunbar and paleontologist David Webb had found evidence of a human presence—a few pieces of stone that looked like tools and a tusk scarred with cut marks that

strongly resembled the work of human butchers—in the sink-hole in the 1990s. Dunbar and Webb had dated the layer the tusk was buried in to around 14,200 years ago.

But Halligan told me that "people were pretty dismissive" about the discovery. "There's a long tradition in archaeology of inviting experts out to see your site while you excavate it," she explained. "If you find something really controversial, you have people come out and look at it. But unlike other sites, because this one is underwater, most archaeologists couldn't visit it to assess it themselves" (1).

Many of Halligan's methods are similar to those of her col-leagues who excavate aboveground—she excavates a single geolog-ical layer at a time, carefully scraping the earth away with a trowel, documenting details of the stratigraphy and artifacts by hand and camera. But in her case, she also happens to be doing this under-water, in scuba gear, which means she can only work for short peri-ods of time before she switches places with a colleague for safety. The sediment she trowels away from each layer isn't collected into a dustpan for screening but sucked by a water dredge through a large hose. "It looks much like your average dryer hose, except reinforced so it's not so bendy," Halligan explained. The troweled sediment is deposited onto a screen that floats on the surface of the water, where other divers carefully look through the dirt as it is deposited, searching for artifacts and fossils from each layer.

It was deep under the water, in geological layers dating to 14,550 years ago, that Halligan found the stones: a fragment of a bifacially flaked knife, along with a flake. A bit farther away, she found additional flakes, in layers dating to 14,200–14,550 years before present.

Even Halligan will admit that these few scraps of rocks can, at first glance, seem quite unimpressive—a single's day excavation at a typical terrestrial archaeological site would probably turn up more artifacts than this. But these little rocks and the story that they tell hold tremendous significance in the context of this site. They tell us that people had visited this insignificant little pond to butcher a mastodon long before Clovis tools appeared in North America.

"Once you get down to the poop levels, it's completely pleasant to dig," Halligan told me about the site, which has come to be known as Page-Ladson by archaeologists. Above those layers there's lots of clay and silt, making it hard to see as you excavate. "But when you get down to the poop, it's just hay and sand and bones and rock."

The "hay" Halligan refers to is the remains of mastodon digesta: cypress sticks, a thorny tree that grew near the pond that the creatures loved to eat, as well as grapes and gourds that the mastodons gorged on. Halligan's enthusiasm about the poop layers is easy to understand; they provide a wealth of information. Poop layers are "super easy to see stuff in, easy to keep a nice straight wall," she told me, "and *every single one* of those sticks is datable. When we found the biface we were able to take samples and dates from all around it.

"We dated the crap out of our walls," she laughed.

Halligan, Michael Waters, and their team were able to obtain over 200 radiocarbon dates, some from sticks lying right next to the stone tool. Together these dates showed that the layers at the site were all in "good stratigraphic order," meaning that each layer was older than the one above it and younger than the one below it.

A stone knife and flakes, a butchered mastodon, an unimpeachable stratigraphy, numerous radiocarbon dates that are all consistent with each other... this is about as convincing as it gets for a pre-Clovis site, short of hearth or human remains. But what exactly makes this discovery meaningful?

Page-Ladson is incredibly important to us precisely because it *wasn't* important. This was just a tiny pond in the middle of nowhere 14,500 years ago, a nice place for big animals to come have a drink and a rest in the middle of the hot savanna.

Archaeologically, Page-Ladson is a very sparse site. There's no evidence of humans living in the area, no evidence of sustained activity or periodic visits to the site. Future excavations may change the way we interpret the site, but right now the flakes and broken knife seem to have been from a one-time visit, a scenario perhaps not unlike the one I gave at the beginning of this chapter. (This sparseness makes some archaeologists look askance at the site, wanting more evidence to be convinced that it's legitimate.)

The sparseness of Page-Ladson, the isolation of it, is *not* what archaeologists expect to see when a group of people moves across a landscape for the first time. Instead, the pre-Clovis peoples of Page-Ladson seem to be "settled in" to the region (2).

Page-Ladson does more than indicate people were in the Americas long before 13,000 years ago—that is pretty well accepted at this point. It helps to reconcile the genetic and archaeological records by showing that people had already been well established across the Americas thousands of years before Clovis. Florida is a long way from either of the two potential paths that humans might have taken to enter the Americas— through the interior along the margins of the Rocky Mountains or along the West Coast. It would have taken time for people to have gotten there and learned the geography of the region and the distribution of its resources well enough to have been living so far inland.

In this chapter and the next, we will explore what the genetics says about how humans got past the ice wall and peopled the continent. We'll be covering the history of the Americas from roughly 20,000 years ago through about 10,000 years ago in this chapter, and the later peopling events (after 10,000 years ago) in the next chapter.

At this point, I think it's important to pause and remind ourselves once again that there are very different perspectives on this period, depending on which kind of evidence you prioritize. People who prioritize archaeological evidence from the Page-Ladson site and accept as valid sites like Monte Verde II in Chile, Paisley Caves in Oregon, and the early sites along Buttermilk Creek in central Texas take the perspective that there was a pre-Clovis presence of people in the Americas as early as 16,000 years ago, certainly by 15,000 years ago, and possibly as early as 20,000–30,000 years ago. The genetic story presented in this chapter is interpreted according to this model. However, archaeologists (and some geneticists) who do not accept these sites as valid or as traces of the early ancestors of Native Americans* will take issue with the model I present in this chapter.

---

* For example, they may represent early peoples in the Americas, but those peoples were not the genetic ancestors of present-day Native Americans.

Their interpretation, which we examined in chapter 3, hinges on Swan Point and later (~14,000 years before present) sites in interior Alaska that show distinct cultural linkages to Siberia and suggest a much later origin of peoples in the Americas.

People who prioritize traditional knowledge (including histories that have been in existence for hundreds, if not thousands, of years) may find points of agreement between the genetics histories and their own, or they may find complete incompatibility between these knowledge systems.

I am skeptical that we will ever come to a perfect agreement among all people curious about the peopling of the Americas, but then again, I don't think that such unity is required in order for us to appreciate the past. The forest of history is healthier and more beautiful for having many different kinds of trees.

About 17,000 Years Ago, the Western Coast of Alaska

As the ice sheets began to melt, the First Peoples expanded southward. This expansion left very clear imprints in the genomes of their descendants.

Mitochondrial lineages show us that after the LGM, people were suddenly and rapidly spreading out, and their populations were growing enormously—about 60-fold between about 16,000 and 13,000 years ago (3). This population explosion is exactly what we expect to see in the genetic record when people move into new territories, where resources are far less limited, there is no competition from other people, and the game animals have no natural fear of humans, having never seen them before.

The story this rather dry genetic evidence reveals is breathtaking when you stop to think about it: a small group of people survived one of the deadliest climate episodes in all of human evolutionary history through a combination of luck and ingenuity. They established themselves in a homeland, from which their descendants—hoping to make a new and better life for themselves—ventured out to explore. These descendants found new lands beyond their wildest expectations, entire continents (possibly) devoid of people, lands to which they quickly adapted and developed deep ties. These ties persisted through millennia into the present day and have not been severed despite climatic challenges and the brutality of colonialism, occupation, and genocide.

But it was the nuclear genome from a small child—who himself did not have any descendants—that gave us the greatest insight into this process.

MITOCHONDRIAL MODELS OF DISPERSAL

Before whole genomes were accessible to give us the details discussed in this chapter, different models for dispersal out of Beringia were proposed based only on mitochondrial DNA. The very earliest studies proposed that each mitochondrial haplogroup—A, B, C, D, and X—migrated separately into the Americas. That was quickly debunked when it was demonstrated that all five haplogroups had (roughly) similar coalescence dates and were present in ancient populations.

Other models proposed a single migration, or three waves of migration (as we discussed in evaluating Greenberg's three-wave model). One model suggested that a migration down the Pacific coast brought people belonging to lineages from haplogroups A, B, C, and D between around 20,000 and 15,000 years ago, and a later migration (containing people belonging to haplogroup X) followed down the ice-free corridor once it was available. Another supporting piece of data for the dual migration model rested upon the geographic distribution of two rare mitochondrial DNA lineages, D4h3a and X2a, which were seen in Pacific coastal regions and northeastern North America, respectively. The finding of a basal X2a lineage in the 8,000-year-old Ancient One/Kennewick Man from Washington seriously undermined this model (4).

12,600 Years Ago, South-Central Montana

Another child's death, another site report that presents the facts in a detached way. But I imagine the Anzick site like this:

Like all their children, the two-year-old boy was treasured by his people. When he died suddenly and inexplicably, their grief was incalculable. His loss would be felt every day. To honor him, they buried him underneath a rockshelter with great care and love, sprinkling his body with red ochre. Everyone in the community contributed to the toolkit that he would take with him into the afterlife: Some placed carefully flaked finished tools—projectile points, knives, and scrapers for hides—others left the cores that he would need to make new ones. His parents placed carved elk bone rods, heirlooms that had been in their families for centuries, into the grave to mark his connection to their ancestors. They sprinkled these with red ochre, too.

This burial site was honored by their descendants for generations, who paid their respects to the boy every time they passed it. Two thousand

years later, when another boy was suddenly taken from his family, they
derived some comfort by burying him close to their ancient ancestor for
protection.

The graves of these two children were found accidentally by construction workers in 1968. Because they were found on private land, their remains were not under the purview of the law that requires consultation and repatriation (if requested) with affiliated tribes.* Nevertheless, after the genome of the two-year-old had been sequenced, researchers (including Sarah Anzick, a member of the landowning family who had done some of the research, and Shane Doyle, a historian and member of the Crow tribe) consulted with the Indigenous peoples of Montana, including the Blackfeet, Confederated Salish, and Kootenai tribes, the Gros Ventre tribes, the Sioux and Assiniboine tribes, the Crow tribe, and the Northern Cheyenne tribes. The tribes agreed that the children should be reburied in a safe place near their original graves, and their wishes were followed shortly after the publication of the study.

The children's names are unknown, but they are referred to by archaeologists as Anzick-1 (the two-year-old) and Anzick-2 (the seven- or eight-year-old who was buried there later). Anzick-1 was special not only to his parents and relatives (both in the past and across time), but also to the scientific community across the world. His remains were dated to between 12,707 and 12,556 years ago, making him the oldest known person in the Americas—the only person who lived during the Clovis period whose remains are known to have survived to the present day. His genome was also the first ancient Native American genome

---

* The Native American Graves Repatriation and Protection Act, or NAGPRA, does not apply to human remains found on privately owned lands. We will discuss NAGPRA in chapter 9.

to have been completely sequenced, and it has given us important insights into the First Peoples' movements into the Americas.

Anzick-1's complete nuclear genome—and those from additional ancient individuals that were sequenced in later years—shows us that shortly after the LGM the family tree of the First Peoples split into two major (and one minor) branches.

Dispersal of First Peoples into the Americas

The minor branch, which diverged between 21,000–16,000 years ago, is currently represented by a single genome from a woman who lived on the Fraser Plateau in British Columbia—known as the Big Bar Lake site to archaeologists—about 5,600 years ago. The fact that her lineage split before the two other major branches may reflect the divergence of her ancestors from other First Peoples as they were moving southward out of Alaska.

One major branch, which included Anzick and his relatives, became the ancestors of many Native peoples of the present-day United States and everywhere south of that. This branch is referred to by geneticists as SNA (Southern Native Americans). The other branch, which is ancestral to populations of northern North America, including the Algonquian, Salishan, Tsimshian, and Na-Diné, is referred to by geneticists as NNA (Northern Native Americans) (5).

This split between NNA and SNA branches tell us a lot about the initial peopling of the Americas. For one thing, most genetic evidence indicates that the split took place south of the ice sheets, because representatives of the Ancient Beringians (Trail Creek Cave and Upward Sun River) are equally related to members of the NNA and SNA groups. If those groups had split before they left Alaska, it's likely that one or both groups would have intermarried with the ancient Beringians, resulting in ancient Beringians being more closely related to one branch or the other.

We also see confirmation of this split and its timing from the mitochondrial genomes of dogs.

Dog mitochondrial genomes rapidly diversify into the four lineages found in ancient North American dogs* at nearly the exact same time as the NNA/SNA split: about 15,000 years ago. With the caveat that these mitochondrial data show us only a small fraction of dog population histories in the Americas—the edge pieces of the puzzle—the radiation of dog lineages that mirrors human lineages is nevertheless extremely strong evidence for this model (6).

---

* Sadly, genetic studies of present-day dogs in the Americas show that the original dogs (First Dogs?) are all but extinct. Of all dogs sampled, only a few (including a chihuahua) showed any ancestry from the First Dogs. Population history models show that they were largely replaced by dogs brought over from Europe; they may have been wiped out by introduced diseases, hunting, deliberate breeding, or a combination of these practices. There is a lot of exciting work currently being done in the field of ancient dog DNA, and this is a topic to watch with interest.

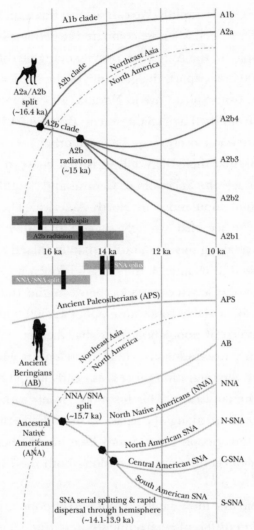

Dogs (inferred from mitochondria) and humans (inferred from whole nuclear genomes) during the Upper Paleolithic. Redrawn from Perri et al., 2021.

Peopling South America

Following the split between the NNA and SNA branches, people belonging to the SNA clade dispersed throughout North and

South America very rapidly. We can see just how rapid this movement must have been when we compare the genomes of the most ancient peoples in the Americas. Despite being on different continents, 6,000 miles apart, the genomes of the Anzick-1 child, an ancient man from Spirit Cave in Nevada (10,700 years ago), and five people from the Lagoa Santa site in Brazil (~10,400 to 9,800 years ago) are very closely related to each other.

The story that their DNA tells us is that between 15,000 and 13,000 years ago the ancestors of Central and South Americans diverged from populations in North America. There are two pieces of evidence that strongly suggest that their movement southward was along the coast, rather than by inland routes. First, as we discussed in chapter 3, the coast was open by 16,000 years ago, whereas the ice-free corridor between the two ice sheets probably wasn't a viable route until about 12,500 years ago. Second, the pattern of population splits that the genomes reveal is so fast—nearly instantaneous—that the scientists who analyzed them likened the migration process as nearly jumping over large regions of the landscape. This fits more closely with southward migration by boat along the coast than with overland migration. By the time people got to South America, via the Isthmus of Panama, they may have expanded along both the east and west coasts.

This rapid first movement was followed by population growth, "settling in" to different environments, and gradual expansions. It was also followed by other significant migrations. After about 9,000 years ago, a group of people from Central America—ancestral to the present-day Mixe from the Mexican state of Oaxaca—spread throughout South America and mingled with all the populations there. They may also have migrated northward as well, as the genomes of people buried in the Lovelock

Cave in Nevada (1,950 to 600 years ago) show us. But as is typical in scientific research, this finding only raises more questions. What caused this movement? And how did traces of Unsampled Population A come to the Mixe genomes about 8,700 years ago? We don't have answers for these questions yet—we are only at the beginning of understanding the complexities of these histories.

Population Y

South American genomes present an additional puzzle, one that was first revealed in 2016 when geneticist Pontus Skoglund and his colleagues analyzed the relationships among ancient and contemporary Native Americans, including South Americans. When they looked at this expanded dataset, they noticed something very odd. The genomes from these populations did not descend from a homogenous group. Probing further, they found that a small number of contemporary populations in the Amazon region—Surui, Karatiana, Xavante, Piapoco, Guarani—shared a small but significant number of alleles with contemporary Australasian populations, including Indigenous Australians, New Guineans, Papuans, and the Onge from the Andaman Islands (7).

"It was one of those results that gives you a combination of pause and excitement for the rest of the day," Skoglund told me in an email. "There are many leads that are wrong in science though, so I knew that it had to be confirmed stringently, and not published right away. With David Reich and Nick Patterson, I worked for about 1.5 years to twist and turn the result, to exhaust the possibility that it was some type of artifact, and to try other methods of detection that could provide independent lines of evidence."

Skoglund and his colleagues called the hypothetical population from whom this ancestry came the Ypikuéra population (or Population Y for short) in recognition of the detection of its genetic legacy in Tupi-speaking populations.

Researchers trying to interpret the patterns of how Population Y ancestry are distributed are still trying to answer some basic questions. If this is a real signal, where did this ancestry come from, and when did it arrive? We can immediately rule out post-European contact gene flow from African, European, or Polynesian groups as a source. Whatever event that produced this signal of shared ancestry between Amazonian and Australasian populations had to have taken place prior to European colonialism beginning in 1492. More research has shown that the signal appears in human genomes from South America as early as 10,400 years ago, and it is broadly distributed across South America in populations along the west coast as well as Amazonian peoples (8).

The tempting interpretation is to assume that Southeast Asians sailed to South America and intermarried with people already present in the continents. If it were true, we would expect to see this ancestry distributed in a diffuse wave from the west coast, with higher proportions of it in populations near the place of original contact (wherever that hypothetically occurred) and less of it in populations farther away. But that is not what we see.

First, the overall proportion of the genomes of contemporary South Americans with this ancestry is quite low, and people within the same populations who carry it have extremely variable amounts of it. It is also quite old, and it seems to predate the split between those populations who lived along the coast and those who eventually moved into the Amazonian region. This is not the genetic pattern we would expect to see if there had

216

been a post-LGM migration across the Pacific from Australia and Melanesia.

In addition, the same genetic signal was found in an individual from the Tianyuan Cave site in China dated to 40,000 years ago; he was related to Amazonian populations and Australasian populations. To Pontus Skoglund and other researchers, the most likely explanation for these results is not that there was a trans-Pacific migration, but instead that there was once an ancient population in mainland Asia that contributed ancestry to both contemporary Australasians and the ancestors of the First Peoples before they left Beringia. Some of the First Peoples who migrated down the west coast would have carried this ancestry with them, as well as those who populated the Amazon. The patchy distribution of this ancestry today reflects the influence of evolutionary forces—like random genetic drift or the influx of new alleles from later migrations—acting unevenly on families and populations (9).

Another possibility is that people with Population Y ancestry were present in South America before the First Peoples arrived (see the "Archaeological Evidence of Population Y?" sidebar). Only additional ancient genomes will allow us to distinguish between these two possibilities...or perhaps reveal others that remain unforseen.

ARCHAEOLOGICAL EVIDENCE OF POPULATION Y?

In July 2020, a team of archaeologists led by Ciprian Ardelean announced that they had found evidence of stone tools in layers dating to between 32,000 and 25,000 years ago in the Chiquihuite Cave in Mexico. These purported tools are simple flaked limestone rocks quite unlike anything seen from any other site in the Americas. They are extremely controversial among archaeologists for several reasons. First, they're not clearly worked projectile points or knives, as we see at pre-Clovis sites elsewhere; they're more ambiguous, akin to the simple stone tools made by early human ancestors in Africa. This raises the question that was the downfall of the Calico Hill Early Man site: Were they actually created by humans, or by natural processes? In other words, were these artifacts or "geofacts"? Caves are very active geological places, with rocks falling from overhangs and ceilings quite often—a phenomenon called breakdown. It is very easy, my archaeologist colleagues tell me, to mistake rocks flaked as the result of breakdown for ones deliberately shaped, especially if the "tools" are made of limestone found naturally in the cave (as the Chiquihuite ones were). If they were made of obsidian or some other high-quality toolstone that had been brought to the site from elsewhere, that tends to argue strongly in favor of human action. This is a serious and reasonable critique on the part of archaeologists. I don't consider it anti-pre-Clovis bias, because many of the people making this critique are very open to the existence of pre-Clovis sites. If we accept these stones as tools, then do we also have to accept all pre-Clovis sites with stones flaked in similar ways, such as Calico Hill? Some archaeologists would say yes: All of these sites are valid, and our standards for accepting the validity of a site as a site need to be loosened. Other archaeologists (probably most) say no: We need to uniformly apply a rigorous standard for accepting any and all sites, and Chiquihuite Cave falls short of that.

I am open to the possibility that people were here before the LGM, and I would be delighted to see convincing evidence of it.

Another possible candidate for a Population Y site is White Sands Locality 2. This site, which I described at the beginning of the book, does not suffer the same problems of ambiguity that many other early sites do: The footprints are numerous and unquestionably human. It's too early to know whether archaeologists will raise serious objections to the 23,000- to 21,000-year-old dates estimated for the footprints, but if they hold true, this site will need to be integrated into the increasingly complex puzzle of the peopling of the Americas.

Let's explore this question from a genetics perspective. Let's assume for the sake of argument that these earliest sites are valid and there were humans in the Americas before the LGM. Can this be reconciled with the genetics?

There are, as I see it, three ways that it could be. The first is if the pre-LGM populations simply didn't leave a detectable genetic legacy. When the First Peoples moved south of the ice sheets, they established populations across North and South America that gave rise to Native Americans, but there was either no mixing between them and the pre-LGM population, or so little gene flow that any trace of pre-LGM ancestry eventually disappeared. This is entirely possible; remember that the same thing happened with the Vikings at the L'Ainse aux Meadows site.

The second possibility is that this pre-LGM population is the source of the Population Y ancestry that we see in the genomes of Amazonian populations. This genetic signal has been a puzzle to researchers because it has no obvious origin. In his book *Who We Are and How We Got Here*, geneticist David Reich suggested two possible scenarios for Population Y: The first is that it was present in a subgroup of the initial Beringian ancestors of Native Americans and was inherited by some populations of First Peoples and not others. He likened it to a "tracer dye" of ancestry. The second possibility, he suggested, was that it was present in a group of people already in the Americas prior to the First Peoples' migration south of the ice sheets around 17,000 to 16,000 years ago. If so, then genetics indicates that they were displaced by the ancestors of Native Americans everywhere except in South America, where they mixed with the First Peoples. In that case, the early sites could have been made by Population Y.

I put both of these scenarios before Pontus Skoglund, the researcher who first discovered Population Y and knows more about it than anyone else. He agreed that if the sites are real, then either scenario—a group unknown to us genetically because they contributed so little to the genomes of Native Americans, or Population Y—could account for the discontinuity between the genetics and archaeological record.

The third possibility is that there was a pre-LGM population that did leave a detectable genetic legacy in descendant populations, but we just haven't found it yet. Our
knowledge of genetic variation in Native American populations is far from complete, especially in North America (10).

The peopling of the Americas did not end when humans reached
South America. The last stages in this epic story occurred with the
migration of populations into two regions that presented totally different ecological challenges: the North American Arctic and the
Caribbean. We will explore these histories in the next chapter.

Chapter 8

A thousand years after his ancestors first came to their home-land, a man stood on the coast of the Arctic Ocean looking out across the sea ice in the direction of the lands they left behind. In that particular moment, however, the man was far less con-cerned with the history of Thule migration than he was with the polar bear walking directly toward us.

This was not by accident; the man was our site's Bear Guard.* He had allowed me to drive with him up the coast a good dis-tance to intercept the polar bear before it could reach our team of archaeologists. Unfortunately for us, the Bear Guard's attempt to intimidate the animal by standing on the seat of his four-wheeler, with hands raised above his head in imitation of another, bigger bear, failed to convince this particular predator—one of the larg-est bear species in the world. As it stalked closer, the Bear Guard lowered his binoculars and sat back down on the four-wheeler.

"Time to go."

I agreed wholeheartedly.

He reversed the four-wheeler and began to accelerate along the coastline in the opposite direction of the archaeologists'

---

* I have withheld his name and refer to him as the Bear Guard out of respect for his privacy.

worksite. The bear likewise changed directions and trailed after us.

"He's hunting us."

I didn't trust myself to reply. I was simultaneously impressed by the casualness of his tone and somewhat guilty about what my mother would say if she knew I was "helping" the Bear Guard, instead of staying back with the archaeologists, where it was safe. But we hadn't found any burials yet that day, and the ride-along was a way to feel useful while the high school students dug shovel test pits.

The high schoolers were digging these shovel test pits as part of a critical collaboration between the residents of Utqiaġvik and an archaeological research team of graduate students and professionals led by Anne Jensen, an archaeologist who lives in Utqiaġvik.

The Iñupiat have lived along the North Slope of Alaska for almost a thousand years. The site we were excavating, called Nuvuk, contained a cemetery that held relatives and ancestors from many members of the nearby town Utqiaġvik (known as Barrow when I was there), including of the Bear Guard and the high school students helping with the dig.

Nuvuk was originally a village located on the very northernmost point of the Point Barrow spit, which extends out between the Chuckchi Sea and the Beaufort Sea. It has been occupied almost continuously for over a thousand years, first by a group known to archaeologists as the Paleo-Inuit, and then by Neo-Inuit (or simply Inuit) ancestors beginning about 800 CE.

The Nuvugmiut (people of Nuvuk) relocated their village southward at least once before the 19th century, as the ocean storms eroded their northern coast, bringing the sea increasingly closer to their homes. Gradually, more and more Nuvugmiut moved to Utqiaġvik, and it surpassed Nuvuk in both size and population by about the 19th century after establishing a hospital, a school, and a Christian church. Nuvuk was not completely abandoned until the mid-20th century; several elders living in Utqiaġvik had grown up in the town (1). But by the time I came there to work, the only thing that remained of the town on the surface of the ground was an expanse of gravel interrupted briefly by scattered animal bones, little patches of grass, and some lingering dwarf willows. The presence of vegetation, however slight, is an important clue, as it often signals former areas of human activity that had resulted in extra nutrients diffused into the gravel: middens and burials.

The close links between Utqiaġvik and Nuvuk made the residents of Utqiaġvik understandably concerned about the fate of the cemetery on the edge of the coast; the pressure from storms intensified by climate change had rapidly increased the coast's erosion rate. Ancestors' remains, at rest for nearly a thousand years, were now falling into the Arctic Ocean. Arctic warming is thinning and melting regional sea ice, extending the ice-free season. Without the protection of the sea ice, the big storm waves erode the coastline much more rapidly (2). The elders of the community made the decision to locate all the unmarked graves and transfer the remains to another cemetery at a safe distance inland. With support from the community and the National Science Foundation, Anne Jensen enlisted the efforts of a group of Utqiaġvik high school students to conduct the excavation and analysis of their ancestors, providing the students with both an employment opportunity and a chance to learn archaeological and laboratory skills (3).

In addition to the archaeological study, the elders also gave permission for another approach to be used to gain more insight into their ancestors' history: genetics. This is where I was lucky enough to enter the project. Although learning how to be a bear guard was an unexpected bonus, the real reason I came to Alaska was to help my postdoctoral advisor, Dennis O'Rourke (now my senior colleague and the chair of the Department of Anthropology at the University of Kansas), in field sampling the ancestors' remains as they were found.

The elders of Utqiaġvik had agreed to allow Dennis and Anne to take samples for genetic testing, with the stipulations that samples be taken as minimally as possible, that any other studies of the ancestors' remains would be conducted within Utqiaġvik, and that the remains would be buried quickly.

I found these guidelines to be far more liberating than restrictive. The unethical history associated with the early archaeological collections and the discoveries made from human remains in the Americas is a lasting legacy, especially in Alaska where 20th century physical anthropologists dug up contemporary cemeteries of Native peoples to extract their remains. The opportunity to work within an explicitly stated framework, composed by the descendants of the peoples I was hoping to learn from, made it easy to do our scientific research on *their* terms.

Once a burial was confirmed under a student's shovel test pit, I instantly abandoned amateur bear watching and began a methodical sequence of protocols. Before approaching the grave, I would don a face mask, gloves, and sleeve guards to cover any exposed portion of my skin (though admittedly there wasn't much I cared to expose; the Alaskan coast, even in the summer, commands quite a few layers of winter clothing). I would then painstakingly excavate gravel and dirt away from the chest area of the burial, scrub my gloved hands vigorously with bleach, and gently remove a small

226

fragment from the ancestor's remains. I immediately placed the fragment into a sterile sample bag and sealed it tight. No one but myself handled those bone samples until they were deep within the sterile recesses of an ancient DNA laboratory 2,600 miles away at the University of Utah in Salt Lake City, where they were then removed from their bags by an approved graduate student, also swathed in protective, bleach-soaked clothing. Results of the research were discussed with the descendants before they were published, and Jensen continues to live and work in the community.

Nuvuk, like the story of Shuká Káa, is a story of collaboration. But the early days of Arctic archaeology were marked by the opposite approach: exploitative research, removal of remains without regard for the concerns of their descendants, and extraction of knowledge without benefit to communities. The early days of genetics research in the Arctic were similar; though guided by regulations on research with human subjects, there was a lack of engagement with communities' concerns, little reciprocity, and very little sensitivity in reporting research results (4).

Today, close partnerships exist between researchers and community elders. The involvement of the Utqiaġvikmiut in rescuing and studying their ancestors was an important collective endeavor that fostered a productive and respectful research environment. We (especially Jensen) were able to incorporate oral histories and community interpretations into our work.*

----

* I and many of my colleagues feel that the respectful joining of Indigenous knowledge and genetics is an approach that can enhance scientific inquiry and foster a stronger atmosphere of collaboration and trust between communities and scientists. As Kat Milligan-Myhre, an Iñupiaq microbiologist at the University of Connecticut, noted in a 2018 conversation on Twitter, a formal method for citing Indigenous knowledge and nonacademic research partners in academic papers would be extremely helpful to further this aim. See twitter.com/Napaaqtuk/status/1030178797872508928.

Because of these collaborative efforts, we were able to learn a lot from the DNA of the ancestors buried at the site and from their living descendants as well. These genomes—and others across the Americas—tell us a lot about the last stages of the initial peopling of the Americas.

The Peopling of the Arctic

As we discussed in chapter 3, the North American Arctic bookends the story of the peopling of the Americas. During the LGM the majority of the Arctic, including Greenland, Canada, the Aleutian Islands, the Alaska Peninsula, and the southern coast of Alaska, was covered by glacial ice. The distribution of this glacial ice affected the geographic choices people made during LGM. They could no more easily have gone into Canada and Greenland than they could have gone south; each way was blocked by ice. Coupled with the extreme environments of these regions, this meant that much of the lands north of 66° 32' N (the Arctic Circle) weren't populated until after the rest of the Americas. As we have previously learned, though people were living throughout Alaska around 14,000 years ago, they did not reach the Aleutian Islands until 9,000 years ago, and the coastal and interior regions of Canada and Greenland until about 5,000 years ago.

The Paleo-Inuit

The first peoples to live at the Nuvuk site introduced at the beginning of this chapter belonged to what archaeologists categorize as a part of the Paleo-Inuit tradition, which extended across the Arctic from Alaska to Greenland (5). The Paleo-Inuit were highly mobile hunter-gatherers who had adapted culturally and physically to the Arctic's extreme environment. Although they were not the first humans to live in Alaska, they were the first to people the regions above the Arctic Circle.

The Paleo-Inuit might have migrated from Kamchatka, arriving in the western Arctic about 5,500 years ago (approximately 3000 BCE or slightly earlier), and reaching the eastern Arctic around 5,000 to 4,500 years ago. By this point, the land connection between Alaska and Siberia no longer existed. But the proximity of the continents meant that there was frequent contact by peoples on both sides of the Bering Strait.* Some hypothesize that the Paleo-Inuit also moved southward through Alaska and peopled the Alaska Peninsula and Aleutian Islands. Others moved eastward across the Arctic coast through Canada to Greenland. They left a very faint archaeological "footprint" across the Arctic, probably due to their small population sizes. We are fortunate that the Arctic environment is so favorable for preservation, because it enables us to see the contours of this footprint.

The Paleo-Inuit made kayaks and subsisted on a wide variety of marine and terrestrial animals, including birds, fish, fox, caribou, musk oxen, and seals, which they hunted with harpoons, spears, atlatls, and bow and arrows, made with tiny stone blades inserted into shafts. They also made other tiny stone tools—scrapers for working hides into clothing and burins for carving bone—and added bifacially flaked knives, stone lamps, and tiny-eyed needles for tailoring clothing to this toolkit. Arctic

---

* There still is today; visits between Iñupiat and Yup'ik peoples with their friends and relatives are made in winter, by snowmobile, over the sea ice.

archaeologists refer to the carefully crafted tools at Paleo-Inuit sites as belonging to the Arctic Small Tool tradition (in Alaska, it's called the Denbigh Flint complex); this tradition seems to have come from the Neolithic Bel'katchi in Siberia.

The Paleo-Inuit are classified in different ways according to geography. In northern Alaska, the Denbigh Flint complex evolved into traditions referred to successively as Choris, Norton, and Ipiutak; the Ipiutak settlement at Nuvuk lasted from around 330–390 CE. In the Canadian Arctic, Paleo-Inuit sites appear around 3200 BCE and are generally referred to as Pre-Dorset, then successively Early Dorset, Middle Dorset, and Late Dorset. In the northern Arctic archipelago and northern Greenland, Paleo-Inuit sites appear around 2400 BCE, and are called Independence I. In the rest of Greenland, they are referred to as Saqqaq, which appears around 2400 BCE.

There were important cultural, technological, and geographic differences between these traditions. In northern Alaska, the Paleo-Inuit maintained long-distance trading networks across the Bering Sea to Kamchatka, along which they traded meteoric iron, chert from the Brooks Range, art, and cultural traditions. They had a rich and complex pantheon of deities, many related to the natural world. In addition to circular dwellings for their families, they built communal structures called *qargi*, spaced out at intervals along northwestern Alaska, which they may have used for community gatherings, trade, and rituals. It is also possible that they made and lived in snow houses, but this is difficult to confirm, as these structures would not have lasted long.

The Pre-Dorset lived seasonally in small camps—likely made up of hunting bands of several families during some parts of the year and in individual nuclear families during other times.

Their summer houses were generally circular, tent-like structures that are visible today as round arrangements of stones that were used to keep down the skin coverings. They likely lived in different kinds of structures during the winter months, possibly snow houses. Archaeologists have documented numerous changes to the tool technologies, artistic traditions, and houses associated with the Early Dorset through Late Dorset periods.

Saqqaq peoples also lived primarily in nuclear family units within circular tent-like structures, which they anchored by double rows of stones that are still visible today. In some cases, they surrounded their tents with a circular wall. Unlike other Paleo-Inuit groups, they apparently did not make figurines, and only sparsely decorated the objects that have been identified from their sites. The peoples in the High Arctic (Independence I) had a very difficult existence in one of the most challenging environments on Earth. They subsisted on seal, walrus, and birds, with seasonal movements between winter and summer sites near hunting spots. Their circular tent houses were marked by raised central areas, known as "mid-passages," which often contained hearths and cooking refuse and may have divided the house into sections for different activities, possibly segregated by gender (6).

Why did the Paleo-Inuit migrate eastward across the Arctic? It's a natural question, but there is no simple or obvious answer. Many factors contribute to the migration of people from one place to another. People may move because of economic instability, hopeful that a new place might offer an easier life. They may be trying to escape from or resolve social conflict through physical distancing. They may have heard of new resources or opportunities and decide to move to take advantage of them. Or they may have to leave because the animal and plant resources that they depend upon have suddenly become unavailable where they live (as was the case during the LGM).

In the case of the Paleo-Inuit, the archaeological record does not show any evidence to definitively support any one of these factors over another. Competition for limited resources may have driven some people to new territories. However, one hypothesis holds that the Paleo-Inuit may have been following musk oxen herds into the Central Arctic. The Paleo-Inuit were highly mobile, moving seasonally between different regions and repeatedly abandoning—and then reoccupying—large regions. This mobility may have naturally led to expansion and migration into new areas in pursuit of resources (7).

Hair and Ice

Just as it has helped to preserve the archaeological traces and oral histories of ancient Arctic peoples, the cold and dry climate has also kept DNA beautifully preserved. This is exactly what a team of researchers, led by Eske Willerslev at the Centre for GeoGenetics in Copenhagen, Denmark, were counting on when they attempted to extract DNA from a tuft of 4,000-year-old human hair. The hair had come from the Qeqertasussuk site in western Greenland, one of the earliest sites of human occupation in the eastern Arctic. The people who had lived here belonged to a Paleo-Inuit culture called Saqqaq by archaeologists.

Very few human remains have been found from Paleo-Inuit sites, making it all the more impressive that not only was Willerslev's group able to recover mitochondrial DNA from the Saqqaq hair tuft in 2008 but, in their 2010 follow-up paper, they were able to report the sequence of his entire genome—the very first whole genome ever recovered from an ancient human! This ancient person's DNA was so well preserved in the hair samples that the research team recovered enough individual DNA fragments to cover almost 80% of his DNA bases with, on average, 20 individual fragments.

We've learned an enormous amount of information from the Saqqaq man's genome. First, his mitochondrial lineage belonged to haplogroup D2a, which had not been seen in any contemporary Inuit population (although it is present in the Unangax̂ of the Aleutian Islands). When Willerslev's group compared the complete genome from the Saqqaq individual to genomes from other Greenlandic Inuit, they found that they were quite distinct from each other. Instead, the Paleo-Inuit are genetically most similar to Ancient Siberians. This suggests that two different groups of people had existed throughout the North American Arctic: the Paleo-Inuit and the ancestors of contemporary circum-Arctic peoples (8).

The Precontact Inuit Migration

The discovery of two distinct genetic groups in the North American Arctic confirmed something archaeologists had already suspected. Beginning about 800 years ago, a new culture emerged across the Arctic in a wave that spread eastward from Alaska to Greenland over the span of just a few centuries. This culture—sometimes called Thule or Neo-Inuit by archaeologists, although I will be referring to it as ancestral Inuit here—represented a completely different way of life from the Paleo-Inuit, and archaeologists were convinced they must be very different groups of people.

The ancestral Inuit were skilled hunters of whales and other marine mammals. They introduced the dog sled and the *umiaq*, two technologies still used by contemporary Inuit. They built winter houses that extended partially underground. Their tools, clothing, and artwork were entirely different from the Paleo-Inuit's.

These peoples are very clearly the ancestors of contemporary Inuit; the traditional tools, houses, culture attributes, and hunting practices of the Inuit are direct extensions of what can be seen in the archaeological record, and the Inuit's own oral traditions confirm it. Their origins have long been the subject of debate by archaeologists. Societies with Inuit cultural features first appear on both sides of the Bering Strait region; they are referred to as the Old Bering Sea culture (200 BCE to 700 CE), which subsequently gave rise to Punuk (800 to 1200 CE) in northern Alaska and the Bering Strait region, and Birnirk (700–1300 CE) in northern Alaska and Chukotka. The immediate cultural predecessors of the Iñupiat of the North Slope of Alaska, Inuvialuit of present-day Western Canada, and Inuit, become visible in the archaeological record around 1000 CE. Exactly where and how their ancestors' culture first developed originated is still somewhat debated but it seems to have emerged out of Punuk and Birnirk during a period of climatic change. The Thule (a group of ancestral Inuit) migrated from the North Slope across Canada during a warming period that shifted seasonal sea ice distribution and with it the ranges of bowhead whales and other marine mammals. The Thule migration may have been in response to these environmental changes, or it may have been for other reasons (9).

The totality of the archaeological evidence suggested that

the rapid spread of ancestral Inuit culture across the Arctic was because people were migrating along the Arctic coast by boat. Furthermore, shortly after the spread of the ancestral Inuit into Paleo-Inuit regions, the latter disappeared from the archaeological record. Was it a population replacement, or did the Paleo-Inuit join ancestral Inuit communities?

This was a question that the Saqqaq genome seemed to answer decisively: The Paleo-Inuit Saqqaq and the Inuit appeared to be genetically distinct from each other. The ancestral Inuit migration had resulted in a complete population replacement of the Tuniit (as the Dorset are referred to in the Eastern Arctic).

But the story was incomplete. The information had been gathered from a single Paleo-Inuit whole genome, and other temporal and geographic gaps in the Arctic's genetic puzzle still existed. In 2014, Maanasa Raghavan (currently an assistant professor in the Department of Genetics at the University of Chicago) and her colleagues published an extensive study of ancient individuals from Siberia, Canada, and Greenland (10). The geneticists were able to obtain mitochondrial genomes and low-coverage whole genomes from a number of Paleo-Inuit individuals, from Thule and Siberian Birnirk individuals, and from Norse individuals in Greenland. Analyses of these mitochondrial lineages confirmed that haplogroups A2a, A2b, and D4 were found in ancestral Inuit while, haplogroup D2a was found in all Paleo-Inuit individuals. Analyses of whole genomes showed that the ancestral Inuit and Paleo-Inuit were indeed different groups of people. They appeared to have experienced gene flow with each other in the distant past, most likely when ancestors from both groups were still in Siberia.

However, almost nothing was known about the genetic variation of the ancient and contemporary peoples in another region that could also have been a candidate for the origins of ancestral Inuit culture: the Alaskan North Slope, the place where we were sampling the remains of ancient Iñupiat ancestors at Nuvuk.

The Utqiaġvik community was supportive of the project and the questions we were trying to answer. One elder suggested testing DNA from the contemporary inhabitants in order to better understand their history. This was an exciting development, and Dennis O'Rourke's former graduate student Geoff Hayes (now a professor at Northwestern University) turned this into a project called the Genetic Analysis of the Alaskan North Slope (GEANS). I was also fortunate enough to be involved in this project, and in 2015 we published the analysis of mitochondrial DNA from 137 contemporary Iñupiat individuals across the Alaskan North Slope. These mitochondrial linages showed us that while most people belonged to lineages common among Inuit elsewhere, the villages on the North Slope did indeed contain a few instances of mitochondrial lineages atypical for Inuit peoples (11). For example, haplogroup D2a, which has only previously been found east of the Aleutian Islands in ancient Paleo-Inuit, was present in contemporary North Slope villages, but not among the ancient Nuvuk ancestors.

Inuit elders and historians play a crucial role in interpreting archaeological data. But they have important insights into genetic patterns as well. When we published our 2015 study of Iñupiat mitochondrial lineages—first suggested by an elder— we unexpectedly found that people on the North Slope were more closely related to Greenlandic Inuit than Canadian Inuit. One elder in the community commented that this result made sense, because Iñupiat found it easier to converse with people from Greenland than from Canada in their languages. We also found that specific mitochondrial lineages were shared between

236

villages on the North Slope in a pattern that followed the coastline, possibly because traveling by water in the summer and by
dogsled on the sea ice close to shore in the winter was faster than
traveling inland.

In 2019, we joined a consortium of researchers to use
genomic data to examine relationships between the Iñupiat,
Paleo-Inuit, and ancient individuals from Chukotka, East Siberia, the Aleutian Islands, Alaska, and the Canadian Arctic.
This collaboration allowed us an opportunity to take a closer
look at the genetic relationships between the North Slope
Iñupiat and other contemporary and ancient Arctic peoples.
We found that the people of the Aleutian Islands, Yup'ik and
Inuit from Siberia through Greenland, and—surprisingly—
speakers of the Na-Dene language family all shared ancestry from a "proto" Paleo-Inuit source in Siberia. Interestingly,
this did not quite match what another research team investigating the Paleo-Inuit reported in the same week. More
genomes will need to be sequenced to tease apart the exact
relationships between these ancient and contemporary groups
(12).

Our consortium of researchers also found genetic evidence that has implications for the origins of the ancestral
Inuit. We estimated that the population specifically ancestral
to both Inuit and Unangax̂ (peoples of the Aleutian Islands,
also called Aleuts) lived between either 4,900 and 2,700 or
4,900 and 4,400 years ago (depending on the method used)
and had gene flow from NNA populations. Later, the ancestors
of the Inuit—but not the ancestors of the Unangax̂—experienced
gene flow with the ancestors of Chukotko-Kamchatkan populations.

One scenario that may explain these genetic results is that

the ancestors of the Inuit and Unangax̂ lived in the Alaska Peninsula region, where they interacted with NNA populations. This may be visible in the archaeological record as the transition between the Ocean Bay tradition (6,800 to 4,500 years ago) and the Early Kachemak tradition at about 4,000 years ago. Later in time, the ancestors of the Inuit and Yup'ik would have migrated to Chukotka and interacted with populations there around 2,200 years ago and established the Old Bering Sea culture, while the Unangax̂ moved westward across the Aleutian archipelago.

This scenario was first proposed by Don Dumond on the basis of archaeological evidence (though he tried to incorporate then-available biological and linguistic data as well). Although highly speculative at this point, it certainly is interesting and worth further investigation if present-day communities are supportive (13).

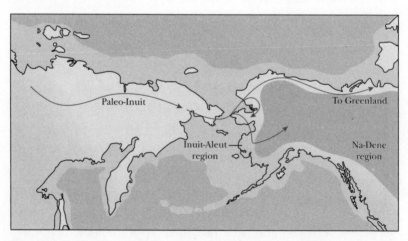

One model for the peopling of the North American Arctic. Illustration inspired by Anne Stone's 2019 *Nature* article, "The Lineages of the First Humans to Reach Northeastern Siberia and the Americas."

ARCTIC ADAPTATIONS IN
NATIVE AMERICANS

Before DNA was available, researchers used teeth in order to reconstruct the relationships between population histories. Teeth are far more reliable markers of ancestry than the skulls that 19th- and early 20th-century physical anthropologists used.

The incisors of Indigenous peoples of the Americas very often have a morphology known as shoveling: an indentation that you can feel as you rub your tongue along the inside surface. This shoveling trait is also found at high frequencies among East Asian populations, but it is uncommon among other groups around the world. It was one of the clues that physical anthropologists used to infer a connection between Native Americans and Asiatic peoples, before it was possible to sequence their DNA.

This shoveling trait is linked to a particular variant (called V370A) in a gene known as EDAR. EDAR has been under strong selection, and V370A is found at very high frequencies in the Indigenous peoples of the Americas—not just in the Arctic but in all populations so far examined.

It doesn't make a lot of sense for natural selection to be working so hard to maintain the shoveling trait—it's not particularly useful for anything, just one of those morphological variants that humans have, like the ability to roll one's tongue. Evolutionary biologist Leslea Hlusko has a hunch that there is much more to the story. She suspects that the shoveling feature is just incidental; V370A influences quite a lot of different aspects of human phenotypes including sweat glands, hair thickness...and the branching of mammary glands.

In the Arctic, as Hlusko explained to me, you can't get enough UV light to make enough vitamin D to stay healthy, even if you stood outside naked (which really isn't an option for very long). Humans have developed effective cultural adaptations to deal with this, including a diet rich in the parts of animals that can give them more than enough vitamin D. However, infants are utterly dependent on their mothers for nutrition, so they are particularly vulnerable when it comes to vitamin D. Perhaps, Hlusko thought, instead of the shoveling trait that's being selected for, it's the role V370A plays in mammary gland branching that's the critical factor in its selection. If V370A leads to increased ability to absorb vitamin D (and perhaps other nutrients as well) via increased mammary ductal branching, then it would explain the intense selection pressure that occurred as the ancestral population was isolated. And it would point to the Arctic as the location for the Beringian Standstill. She is doing more research to test this hypothesis (14).

The Peopling of the Caribbean

The Arctic was one of the last places in the Americas to have been peopled; the other region was an archipelago half a world away in the Caribbean Sea.

Like the North American Arctic, the histories of how people migrated into the Caribbean islands were marked by dynamic population movements, expansions, and adaptations. Genetics offers tremendous potential to reconstruct these histories, but it has been difficult until recently to recover significant amounts of ancient DNA from populations in this region because the hot, humid climate is unfavorable for DNA preservation. Nevertheless, there have been some important studies that have given us an overall picture of the peopling of the Caribbean.

Evidence for the very earliest peoples in the Caribbean is scanty, owing to uneven archaeological research and the likelihood that a number of early sites were submerged by rising sea levels.

The first known traces of a human presence on the island of Trinidad can be dated to as early as 8,000 years ago. Prior to this, the island was connected to South America and accessible via overland travel through the end of the Pleistocene (11,000 years ago). This group has traditionally been referred to as the Ortoiroid culture, and marked by characteristic stone tools at these early sites, but the absence of ceramics.* In recent years, some archaeologists have increasingly questioned the traditional classification systems (Ortiniroid, Casimiroid, and Saladoid) as not accurately reflecting the cultural diversity of the pre-European contact period in this region. I will be presenting newer chronologies, although mentioning the older categories as well to give interested readers an idea of the scope of this debate (15).

Humans reached Cuba and Hispaniola by at least 6,000 years ago. Archaeologists refer to this time as the Lithic Age, and these

---

* There is some evidence for ceramic production during the Archaic age, although mostly from Cuba and the greater Antilles with very little from Trinidad.

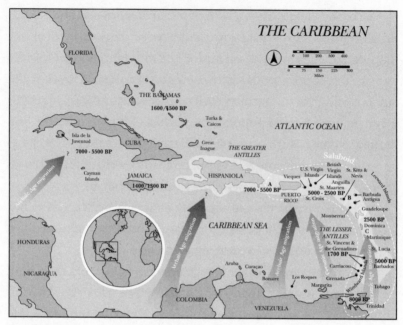

In the Caribbean, Ceramic Age migration was 2,500 years ago, and Archaic Age migration was 6,000–2,500 years ago.

early sites are marked by flaked stone tools (blades and flakes), indicating that they relied upon hunting terrestrial and marine animals and eating wild plants.

Between about 6,000 and 2,500 years ago, a time period called the Archaic Age by archaeologists, sites belonging to the Casimiroid culture begin to appear in the archaeological records of Cuba, Antigua, and other islands.* These sites are characterized by polished stone tools, suggesting the people relied upon fishing as well as terrestrial mammal hunting, and cultivated and ate a wide variety of local plants.

---

* Though, oddly, not Jamaica or the Bahamas. This may be because of poor site preservation, or it could be because it was too difficult to reach these islands.

However, some archaeologists claim that the separate classification of Lithic Age and Archaic Age sites is based upon an inaccurate distinction. Because the two groups of sites have many overlapping features, these archaeologists believe that they should all be classified together. Whether or not their establishment on the islands was due to the migration of different people remains a long-standing question in archaeology.

Regardless of how they are classified, the sites from the Lithic and Archaic groups look very different from those that begin to appear about 2,500 years ago, during what is called the Ceramic Age. Marked by a distinctive style of pottery and indications of an increasing reliance upon agriculture, Ceramic Age sites have been interpreted by many archaeologists as the result of the arrival of a new group of people; a culture historically referred to as Saladoid.* Ceramic Age peoples not only made fine ceramics; they also crafted portable art and sacred objects from minerals such as amethyst and jadeite. They manufactured stone tools and cultivated many plants, some of which they imported from South America, including potatoes, maize, peanuts, chili peppers, and the hallucinogenic beans from the yopo plant *Anadenanthera peregrina.* They likewise relied heavily on marine resources, including fish, crabs, and birds.

Archaeological evidence and early genetics studies connected these peoples to South America, and there have been two major archaeological models for their arrival. In one scenario, they gradually moved northward from the Orinoco River basin of Venezuela, through the Lesser Antilles to Puerto Rico, and eventually expanded westward into Hispaniola and Cuba. The

---

* The previous term was Arawak.

second scenario suggests that they first moved into Puerto Rico, and then expanded southward (16).

Over the last few decades, a number of genetics studies provided us with broad outlines of population history and how the genetic variation in contemporary Caribbean populations was shaped. Then in 2020 and 2021, two large-scale ancient genomic papers offered a sharper, highly detailed picture of the peopling history (17).

Collectively, genetics research shows that in the history of the Caribbean, two major population migrations occurred. The first was an early Archaic Age migration, possibly originating from South or Central America. But far more work needs to be done in order to better understand this migration.

"To be perfectly frank, we have no idea where precisely the first peoples of the Caribbean come from," Maria A. Nieves-Colón, a Puerto Rican anthropological geneticist specializing in precontact history, told me. "The archaeological data suggest connections with both South and Central America, and the genetic data available to date points to ancestries outside of present-day Indigenous American variation. We think it could have been either of these two areas (or perhaps both?), but where within this broad region is a mystery. North America cannot be entirely ruled out either, but there is no archaeological evidence to suggest it as a likely source area, so it's not seriously considered anymore."

A later Ceramic Age migration likely originated from northeastern South America. Genetic affinities between ancient Ceramic Age individuals in Curaçao and those from the Lesser Antilles support the archaeological model for a northward expansion from the Orinoco River basin.

Similarly to the Paleo-Inuit and Neo-Inuit half a world away, as these new peoples spread throughout the Caribbean region, the lineages found in Archaic Age populations dwindled in frequency, so much so that there was eventually a near-complete population turnover in all localities examined—except western Cuba. There, researchers found lineages from the two populations persisting side by side for more than two millennia, reflecting the endurance of two distinct ethnic groups (18).

Somewhat surprisingly, geneticists have found very little evidence of intermarriage between members of the two populations. Only a few individuals from the several hundred that were examined showed ancestry from both Ceramic Age and Archaic Age populations.

Considering how difficult it is to retrieve ancient DNA from hot and humid climates, these whole genome studies and older ancient DNA studies that were focused on single genetic markers (mitochondrial and Y chromosomes) are remarkable technical achievements. Together with DNA from contemporary populations in the region, they help us understand not only the origins of the peoples of the Caribbean but also how their genetic legacy has continued into today.

Ceramic Age peoples were likely ancestral to the Taíno, who were the First Peoples encountered by Columbus and his crews, and other cultural groups who may have been present at the time of contact (19). Brutal colonization practices, as well as the diseases introduced by Europeans, were long thought to have completely eradicated the Taíno. However, genetic studies of contemporary peoples of Puerto Rico have shown the persistence of Indigenous mitochondrial DNA lineages, and Indigenous (Ceramic-related) ancestry persists in present-day Puerto

Rican and Cuban individuals. This finding has been embraced by Indigenous Caribbeans, who celebrate the resilience of their ancestors (20).

Resilience

Although these later peopling events occurred in the Arctic and the Caribbean—vastly different climates, to say the least—they share interesting parallels. Each provides us with a good example of how cultural differences observed in the archaeological record do sometimes reflect the presence of two genetically distinctive groups of people. They also give us an important perspective on how ocean travel presented few barriers to the experienced seafarers in both regions. Inuit traveled back and forth across the Bering Sea quite frequently, and the Caribbean Sea has been characterized as "an aquatic motorway" for early peoples by one team of researchers (21).

Another parallel between the first peoples in the Caribbean and the Arctic is, sadly, the appalling legacy left by contact with "New World" explorers and colonizers. Though the Caribbean was the first place that Columbus and his crews made landfall, and the Arctic was one of the very last places to encounter Europeans, Indigenous peoples in both places suffered terribly from brutal colonization practices and the introduction of new diseases.

The fact that Indigenous culture and identity in both regions perseveres is a testament to the resilience of Native peoples, even in the face of near-apocalyptic conditions.

And right now, we are living in an extraordinary moment: We have the ability to learn a great deal more about the histories of this period because of technological advances made in the field of genetics.

Genetics is a tool that can be added to interdisciplinary work in archaeology, linguistics, and Indigenous historical knowledge, governed by the interests, priorities, and concerns of stakeholder communities. Fortunately, these collaborations are flourishing in both the Arctic and the Caribbean, offering much promise for understanding these histories in the near future.

Although there is a disproportionate amount of research attention to questions regarding the origins of Native Americans and the initial peopling of the Americas, Indigenous histories should not be viewed as limited to the distant past. There are many other important questions to be asked about the thousands of years that followed the initial peopling: how populations settled into their lands; how people adapted, changed, and maintained traditions; how people traveled and encountered each other; how tribes maintained peaceful relations or engaged in conflict. Tribes and communities have extensive knowledge about these histories; some believe that there are ways in which genetics can help reveal or clarify these stories. But unfortunately this potential is marred by a history of shameful behavior of some geneticists and anthropologists toward Indigenous peoples. These challenges—along with ways for addressing them—are what we will explore in the next chapter.

Chapter 9

On July 28, 1996, just 24 days after Shuká Káa was found, two young men stumbled upon the remains of a person eroding out of a riverbank near the city of Kennewick, Washington. The coroner who investigated the remains asked archaeologist Jim Chatters for help in recovering them and determining as much as possible about the individual. Chatters's initial examination of the skeleton's cranium led him to believe that the remains were of a middle-aged man of European descent, although, oddly, he had a stone projectile point embedded in his hip. A sample of bone was sent off for radiocarbon dating, and like the one from Shuká Káa, the sample returned a shocking age: around 9,000 years old. Unlike Shuká Káa, however, this is not a tale of a productive collaboration between scientists and living descendants to learn about the past (1).

The man's original name is not known to us, but some of his present-day descendants, members of the Umatilla tribe in Oregon, call him Oid-p'ma Natitayt or the Ancient One (2). To most archaeologists he is known by another name, Kennewick Man, after the present-day city of Kennewick, Washington, near where his body was found.

Oid-p'ma Natitayt might well have been surprised at how bitterly, 9,000 years after his death, anthropologists have argued about the shape of his head and what it did or did not say about his own ancestry. When his skeleton was first found, he was assumed to be a man of European descent because of the way his skull was shaped. When it became clear that he predated European contact by several thousand years, his long, narrow, and tall skull became the focal point of a controversy. At a press conference, Chatters described him as belonging to the "Caucasoid type," very distinctive from the "Mongoloid type" that "Amerindians" belonged to.

Chatters insisted later that he did not mean to imply that Kennewick Man was "white," but unfortunately the press and the public heard "Caucasian" and "9,000 years old" and this launched breathless speculation about his origins and what he might mean for the history of the Americas (3).

This speculation was not helped by a reconstruction of the Ancient One's face that was directly inspired by Captain Jean-Luc Picard from *Star Trek*. This reconstruction provided visual evidence far more convincing to many people than the objections of anthropologists that there was no conceivable way that Kennewick Man could be European. White supremacists pointed to the reconstruction as evidence contradicting Native sovereignty over American lands. Wild headlines in the news proclaimed that the Kennewick Man's skeleton overturned all paradigms in American history (4).

But did it?

The skulls of the very earliest people found in the Americas—those dating to the late Pleistocene/early Holocene—look distinctive from those of Native Americans from later periods. They tend to be longer and narrower, with faces projecting slightly more forward and their eyes and noses set lower on their faces. This suite of features has been called a Paleo-American (or sometimes Paleo-Indian) morphology. It is seen in the earliest ancestors buried at the Lagoa Santa in Brazil, Kennewick Man, and the very few other individuals dated to the late Pleistocene/ early Holocene across the Americas.

This has led some researchers conducting craniofacial morphology studies of ancient First Peoples to suggest that they derived from two biologically different groups. The first group consisted of the "Paleo-Americans," and because of the age of the human remains with this morphology, they were thought to represent the very first peoples in the Americas. Who exactly they most resembled depends on which researcher you talk to; various studies linked their cranial morphology to those of contemporary sub-Saharan Africans, Europeans, and circum-Pacific populations. Some researchers have suggested that they are the remnants of a very ancient migration (from Southeast Asia, Africa, Europe, etc.) (5).

According to this hypothesis, a second group descended from East Asians (which exhibited a more "Mongoloid"* morphology: flatter face, eyes and nose set higher up in the face, a shorter and wider cranium) was thought to have arrived separately after the Paleo-Americans and was ancestral to all later Native Americans, including present-day peoples.

By now, having read most of this book, you know that genetics does not support this model for human origins on the American continents. Let me be even more explicit: All ancient individuals from whom we have DNA, even those with this Paleo-American morphology, are genetically most closely related to present-day

---

* In some places in the anthropological literature, the 19th-century skull typology that once permeated physical anthropology lingers like the smell of stale cigarette smoke.

Indigenous peoples of the Americas. Even the finding of Population Y ancestry does not lend support for the Paleo-American hypothesis, since many people with this morphology show no trace of Y ancestry (6).

But the Ancient One's body was found before the paleogenomics revolution. Although attempts were made to sequence his DNA, the methodologies back then were not sensitive enough to recover enough DNA to identify him as Native American.

The Ancient One's age led tribes in the area—the Colville, Nez Perce, Umatilla, Wanapan, and Yakima—to claim him as their ancestor under the Native American Graves Protection and Repatriation Act (NAGPRA) and request repatriation of the remains to them by the federal government (he was found on land managed by the Army Corps of Engineers). In September 1996, the Corps halted all research and published a "Notice of Intent to Repatriate Human Remains" as required by NAGPRA. They refused to allow scientists to further study the remains.

From the perspective of the scientists involved in the case, this was a unique opportunity to learn about the life and history of the earliest peoples in the Americas, whose remains are extremely scarce. Because the remains were so ancient, scientists argued, they could not be linked to any single tribe but were potentially ancestral to all peoples of the Americas (or none of them, if you believed that his cranial shape was a reliable marker of ancestry). Therefore, it was critical that he be studied and his remains be treated and stored according to the best known practices for preservation so that they would be available for study by future scientists with new approaches. The refusal of the Corps to allow the remains to be studied—and their decision that the remains were to be returned to local tribes—was unconscionable to these scientists. A group of them sued the Corps to prevent the repatriation of the Ancient One to the tribes, who it was assumed would immediately rebury him.

The Ancient One's body, 9,000 years after his death, became the focal point of a fight between some archaeologists and the Indigenous-led movement (which also included non-Indigenous scientists and bioethicists) for control over their ancestors' remains. The fight would be a hugely significant event in the history of anthropological and Native relationships. Its effects are still felt today (7).

After extensive legal wrangling, the scientists won the case. They were allowed to study the remains and produced a detailed, 670-page book on their research (8). From their work, we learned much about the life of an ancient person from this era.

The Ancient One's body reflected the wear and tear of a hard life. His teeth were worn and damaged from a lifetime of eating abrasive food, possibly dried fish (he had eaten a lot of salmon in his lifetime). He had some small bony growths in his ear canals (called auditory exotoses), which suggests that he had been exposed repeatedly to cold and wet environments. They would have slightly damaged his hearing.

The man was almost certainly a hunter: His spine and joints showed early stages of arthritis and attested to rigorous activity since childhood. His right shoulder resembled that of professional baseball players: extremely developed, damaged with stress fractures, and probably chronically sore from years of throwing a spear using an atlatl (a spearthrower).

Several of his ribs had been broken from a hard blow to his right chest. Was the blow delivered by a kick from an animal he was dispatching? Was it caused by an accident? It's impossible to tell. The ribs healed in a way that shows he was unable to rest comfortably following the injury—a testament to a life of activity. Another event early in life caused a small depression fracture to his skull. It had long since healed. Sometime when he was a teenager, the man had been impaled by a spear, possibly a hunting accident or violent attack. The spear had been thrown with such force that its stone tip broke off in the man's hip. He'd been lucky—the spear missed his organs and did minimal damage (although he would have taken a long time to recover from the injury, and the stone point would remain in his bone forever). Or perhaps it wasn't luck, but rather his own skill at hunting that caused him to pivot in time to successfully minimize the damage.*

He was likely between 35 and 40 years old when he died. We don't know what killed him. It's possible that he died from an infection, but he could equally likely have died from something else that left no trace on his bones.

Geneticist Eske Willerslev was eventually allowed to attempt another study of Kennewick Man's DNA and found that, contrary to the story told by his skull, he was closely related to all other Native Americans. Members of the Colville Tribes allowed Willerslev to sample their DNA in order to compare it with the Ancient One's genome; they shared genetic ancestry. The extent to which he is related to other North American groups is unknown as we have very little genetic data from Native Americans in the United States. It's probable that he is related to many different groups in North America. The research didn't establish that he was the ancestor of any tribes to the exclusion of any others, but it did demonstrate that he was Native American and his population related to the members of the Colville Tribes (9). Although repatriation of his

---

* Della Cook, with whom I was lucky enough to study at Indiana University, discusses this idea at length in her evocatively named chapter within the Kennewick Man volume: "The Natural Shocks That Flesh Is Heir To."

remains to the tribes was originally denied under NAGPRA, in 2016 after the genetic study was published Congress passed an act to do so titled "Bring the Ancient One Home." The Ancient One was repatriated to the consortium of claimant tribes on February 17, 2017, and was reburied on February 18 (10).

To some scientists, this repatriation represented an incalculable loss of an important source of evidence about the past. They see remains of this age as belonging to humanity as a whole. As an individual this ancient man is a potential ancestor of so many different peoples, they feel that it is unjust for only a few to assume the responsibility of deciding what is to be done with him.

But the Indigenous peoples of the region were terribly harmed by the delay in reburial and by the ugliness that the lawsuit dredged up. To the members of the tribes of the Columbia River Plateau, the reburial was a necessary and respectful step in the care of their ancestor, finally laying him to rest after a long and painful fight on his behalf.

Kennewick Man brings to the forefront of public discussion complex questions that geneticists, archaeologists, ethicists, and Indigenous peoples grapple with constantly: Who speaks for the dead? In cases of the recently dead, when there are living relatives or direct descendants, this is more clear-cut; obviously their wishes should take priority over those of nonrelated scientists. But even in that scenario, what if you and your cousin disagree about whether your grandmother's remains should be used in scientific research? Whose wishes take precedence? Now extrapolate that backward in time 1,000 years, or 10,000 years. Many tribes are potentially descended from the population to which the Ancient One belonged (we'll sidestep the question of whether he himself was directly ancestral to anyone). Who can give or refuse permission for research on an ancient person if there are potentially many thousands of descendants? Should the default position be one of respecting the wishes of tribal members who claim kinship with him? Or should priority be given to the conducting of scientific research that could potentially give insights into the histories of many peoples?

The troubling history of archaeology in the Americas, which we've discussed throughout this book, overlays these complex questions. When so many ancestors' remains were removed from archaeological sites, often without consulting descendant communities, many tribes feel that they have no reason to trust researchers. And the idea of descendant community that's applied is often based upon a Western notion of "blood relation." Even if an ancestor is not a genetic relative, many Indigenous communities still identify them as such and feel an obligation toward them.

Unfortunately, existing research regulations leave many avenues open for problematic research practices. In the United States, the Common Rule, which underlies ethical oversight of research on humans (governed by Institutional Review Boards, or IRBs), does not apply to ancient humans (11).

NAGPRA, the law that governs human remains and artifacts associated with descendant tribes, does not apply to the remains of ancient people found on private property; a property owner can elect to have genetics research done on such remains without consulting potential descendant communities. This was the case for the Anzick-1 child, although fortunately consultations did take place following the research (12).

Furthermore, NAGPRA does not prevent research on human remains that are deemed to be "culturally unaffiliated." A recent example of problematic research resulting from this loophole occurred when geneticists reported the sequencing of mitochondrial genomes from a number of people buried in Chaco Canyon, New Mexico. Under the law, scientists were not required to consult with any tribes, as the remains were officially designated "unaffiliated." But there are a number of Indigenous Southwestern communities whose oral traditions link them strongly with the ancestors at Chaco, and they were upset that they had not been consulted before the research was conducted. Although it yielded very interesting findings—most notably, the existence of elite maternal lineages that persisted for generations at the site—the knowledge gained in this study and others came at the cost of harm to the communities and further erosion of trust in scientists (13).

Violation of the Sacred

The reluctance of some communities to participate in genetics studies or give permission for ancient DNA research also stems from a complicated and troubled history of biomedical research with Indigenous peoples (14). When I talk to Indigenous peoples about genetics research, they often point to the case of the Havasupai as a reason why research should be viewed with caution.

The Havasu Baaja (People of the Blue Green Water) (15), who belong to the Havasupai tribe, live in one of the most remote and beautiful regions of the Grand Canyon. A small population of about 750 people, they unfortunately have high rates of type 2 diabetes, which causes great suffering within the community. From 1990 to 1994, the tribe permitted researchers from Arizona State University to collect over 200 blood samples in order to investigate potential genetic causes for the high rates of this disease in their community.

At least, that's the purpose for which the Havasupai believed they were donating their blood. In fact, the written consent forms that they signed indicated that they were donating their blood to help researchers "study the causes of behavioral/medical disorders." The broadness of this consent meant that legally, many different kinds of studies could be done with their blood.

Carletta Tilousi, a member of the tribe, attended a PhD dissertation defense in 2003 at ASU. In an interview with the *Phoenix New Times*, she talked about how shocked she had been to hear the graduate student present research he had done using her own DNA: "He spoke about how the DNA of this isolated, intermarried group of people—us—was unique, and how my people had migrated to Arizona from Asia" (16). The tribe's traditions hold that they came from the Grand Canyon region; they did not knowingly consent to genetics research that thus undermined their historical knowledge, cultural identity, and potentially risked their own sovereign claims to their lands. "I started to think, 'How dare this guy challenge our identity with our own blood, DNA,'" Carletta told the *Phoenix New Times*. "Then I remembered when many of us gave blood years ago for a diabetes project. I wondered if this was the same blood."

During the question-and-answer session after the presentation, Carletta identified herself as a member of the tribe and asked the graduate student if he had received permission from her tribe to conduct the research. Upon further investigation, it became clear their DNA had been shared with other labs and used for other kinds of research—including population histories and schizophrenia—not approved by the tribe.

When the university refused to apologize and return the samples, the tribe banished *all* researchers from their lands. Eventually a lawsuit was settled in the tribe's favor, but by then the damage had been done. Other tribes, alarmed by the Havasu-pai's experience with geneticists, have subsequently refused to participate in genetics research (17).

Many Indigenous peoples view hair, blood, and tissues of the body as sacred, and that the disturbance of ancestors' bodies for DNA study is disruptive and harmful. Thus, conducting any genetics research on their ancestors may be simply incompatible with their values.

Furthermore, as exemplified by the Havasupai case, the practice of keeping samples indefinitely, sharing them with other researchers, carelessly losing samples, not regularly engaging with the community to inform them about the progress of research, and using DNA for purposes to which they did not approve (even if consent had been broadly given)—all of this was a deep violation of the trust tribal members placed in the scientists' hands along with their sacred blood.

DNA and Racialization

The concept of race has long been a weapon wielded against tribal sovereignty. In the United States, efforts by the government to "civilize Indians" encompassed the theft of land but also the attempted theft of identity; the suppression of language, removal of children from their families to live at boarding schools that forced assimilation. To these efforts were added state-imposed rules about tribal membership: a definition of who got to be Native American based in some cases on lineal descent from an enrolled tribal member, in others on the concept of "blood quantum."

Today, tribes determine their membership through a variety

of ways—but commercial ancestry testing isn't one of them* (18). "We don't need a swab in our mouth to prove who we are," one person put it (19).

Some people are increasingly wishing to use commercial genetic ancestry testing in order to seek out evidence of Native American ancestry. There are complicated reasons for doing so. Many people, like former Democratic presidential candidate Elizabeth Warren, have family legends of Native Americans in their past; I constantly hear about people's earnest excitement about their "Cherokee great-grandmother," and get questions about how they might be able to use DNA to confirm this story. People have even emailed me their genetic ancestry test results and asked me to interpret them (20)!†

We love having a story that connects us to the past. We love to imagine our ancestors as living, interesting people doing interesting things. I certainly love imagining my great-grandmother, who played the harp in a vaudeville band. I'd like to believe that our love of music and our shared ancestry connects us across the two generations that separate us. This is a normal, healthy, and understandable fantasy. But having "Native American DNA" is not what makes someone Native American (21). You may love your family tradition of a Native American ancestor; you may feel an affinity for your ancestor, as I do for mine. But that is not what makes a person Native American any more than my ancestor

---

* Some tribes do use paternity testing as part of enrollment. This reflects efforts to acknowledge and address the severe disruption caused by governmental and religious programs of enforced assimilation, such as the removal of children from their homes and enrollment in boarding schools or adoption into white families. Paternity testing can be one step in reconnection with Native communities.

† Please don't do this.

makes me a vaudeville harpist. Genetic testing can be a start to establishing a connection with one's Indigenous ancestry, but it can't serve as a substitute for the work of building ties to a community (22).

For Indigenous peoples this is not a minor or abstract issue: Giving legitimacy to the notion that one can claim Native identity via a DNA test or family legend without any connection to present-day tribes is an implicit threat to tribal sovereignty. The illegitimate claiming of Native American identity—and the reaping of benefits designated for minority-owned businesses or other social or educational benefits—is a widespread problem. There's even a name for this broader phenomenon: Pretendian.

Because of widespread interest in documenting Native American ancestry, commercial ancestry testing services are intensely focused on obtaining Native American DNA from tribal members. So far, these attempts have been met with a fair amount of resistance, as have many attempts to enroll tribal members in genomic research studies. The marketing of ancestry tests as "telling you who you are" is not accurate (23). It boils down to this: What makes a person Indigenous is not a subject on which I—or any other non-Native geneticist—can speak with any authority or knowledge.

Vampire Science

I periodically participate in the Summer Internship for Indigenous Peoples in Genomics, an intensive week-long workshop designed to train Indigenous peoples from across the United States in genetics methods and bioethics (there are similar workshops in Canada, New Zealand/Aotearoa, and Australia).

Throughout the workshop, faculty and participants (who include everyone from undergraduate students to postdoctoral researchers to senior tribal leaders) spend hours sequestered inside classrooms and laboratories grappling with complex ethical issues, extracting their own DNA, learning about genetics methods, and discussing how the field of genetics could benefit from the inclusion of Indigenous perspectives.

Many of the programs that compare genomes are complicated to use and can lead non-experts to overly simplistic (or outright incorrect) assumptions about history and race (24). On one occasion that I attended, I had assumed that this issue would be the focus of the discussion, and I was interested in hearing participants' perspectives on it. But I was surprised to hear the conversation going in a very different direction.

I'm uncomfortable using the genomes from Native Americans in this database, one person said.

They weren't collected ethically, another person agreed. *We shouldn't use them. Not even for training.*

Others disagreed. *We should learn what we can from these databases*, a person suggested. *It's important that we learn these methods so that we can carry out the research ourselves.*

They were referring to genomes from several populations— Karatiana, Surui, Colombian, Maya, Akimel O'otham (also called the Pima)—that are publicly available and routinely used by many research papers as representatives for all Native Americans.

These genomes were collected as part of the Human Genome Diversity Project (HGDP), an ambitious international genetics research collaboration that began in the 1990s (25). It, and other major databases of human genomes that were developed after it, like the 1000 Genomes project, the International HapMap Project, the Genographic Project, and the Human Genome Diversity

Project, have given free access to genomics information essential for countless researchers around the world (26). However, some Indigenous peoples find these databases troubling because of their history (27).

HGDP organizers and researchers particularly aimed to collect genetic samples from Indigenous peoples worldwide. The rationale for doing this made sense at first glance: One can't study worldwide human genetic diversity based on a limited geographic sampling. But its intense focus on Indigenous communities and particularly the way the project was initially conceived and implemented led to widespread criticism among Indigenous peoples, bioethicists, and physical anthropologists.

Among the concerns raised was the possibility that researchers might patent genes or otherwise profit from biomedical discoveries using Indigenous peoples' DNA without returning benefits to them. A second issue was the potential for the study to bolster—even if inadvertently—scientific racism with its focus on "unadmixed" populations, language and framing that implied that there was such a thing as "genetic purity." Additionally, the HGDP's rhetoric about sampling "vanishing indigenous peoples" and "isolates of historical interest" was deeply unsettling to the Indigenous peoples described in this way, as it implied that they were relics of the past rather than living members of communities. There was also a significant concern that because project scientists were themselves defining groups for the purposes of genetic sampling, the results from the HGDP would undermine tribal sovereignty and the right to self-definition of their identity. Finally, one of the major problems was the researchers' consent structure, which was designed around obtaining individual consent to participation in the study rather than community consent, as is highly important in many Indigenous groups (28).

In response to concerns raised by tribes regarding consent, the HGDP's North American Committee developed the Model Ethical Protocol for Collecting DNA Samples (1997), which provided a series of rules that all research activities in North America were required to follow. On its surface, this protocol was a good articulation of how genetics research should be conducted in marginalized communities. Among other recommendations for ensuring privacy of participants, returning benefits to participating communities, governing commercial use of samples, and combatting racism, the protocol required community consent and a justification for why researchers wished to include their samples in the HGDP. This protocol also required researchers to "explain both why they concluded consent was appropriate at the levels they chose and why any particular entity was considered a culturally appropriate authority" (29).

By the time the Model Protocol was published, however, the HGDP's reputation among Indigenous authorities was severely damaged; it was even labeled a "Vampire Project" by the World Council of Indigenous Peoples (1993), and the name stuck.

In the end, none of the Indigenous populations in the United States chose to participate in the project, and to this day the HGDP is viewed as a cautionary tale by many tribal leaders (30).

Some Indigenous scientists view the landscape of existing research protections, like the Model Ethical protocol and Memoranda of Agreements between communities and universities as ineffectual, with little or no ability to punish researchers who violate them.

Subsequent efforts to characterize genetic diversity within Native American populations that followed have been viewed with concern, as is reflected by the very few Indigenous communities who have participated in them.

Research that has caused harm to Indigenous communities in the Americas (and elsewhere) has poisoned efforts to understand their histories using genetics. As of this writing, there are very few publicly available genomes from contemporary Indigenous North Americans, and even fewer from ancient North Americans that have been sequenced enough to confidently do detailed population comparisons (31). In comparison, there are thousands of genomes from contemporary Europeans and hundreds of genomes from ancient European populations. This bias is problematic because it excludes Indigenous peoples of the Americas from potential benefits of genomics research (such as genomic medicine, the use of genomics for repatriation claims, or capacity building within communities).

Many of the projects I described above had, on surface examination, good intentions or noble aims. But the ignorance (or, in some cases, outright bigotry) of non-Indigenous researchers has harmed participants, and some Indigenous communities have understandably responded with distrust of researchers. Quite often in my work I have encountered colleagues who are utterly baffled by this lack of trust, by reluctance to participate in—or steadfast opposition to—genetics research. Some people have seized upon the concerns raised by Indigenous communities as evidence that Native peoples are "anti-science." There have been recent accusations that efforts to repatriate ancestors and right the balance of power between communities and non-Indigenous scientists are hostage to "traditional American Indian animistic religions" (32). Far too many researchers, unaware of the history of their own discipline, mistake opposition to research for an anti-science attitude, and that perspective can diffuse into the general public.

But "anti-science" is not a fair characterization of the

communities' perspectives. The scientists involved may not see it, but such accusations are only the latest episodes in a long colonial tradition of disregarding the views and opinions of Native Americans.

The scarcity of genomes from Native North Americans and uneven geographic sampling of populations within the Americas has also led to an intensively competitive environment in paleogenomics. This environment incentivizes projects that sequence as many ancient North American genomes as fast as possible over slower, more engaged research. This, in turn, reinforces the practice of science without consultation; researchers feel that they don't have the time (or the expertise) to invest in developing long-term trust relationships with communities and will instead go for the easiest samples to acquire: those "unaffiliated" remains in museums and universities, often the same collections built by Hrdlička and his colleagues. Or they will give up on doing genetics in North America and focus on regions where consent is more clear-cut or easier to obtain.

This situation is not going to be resolved by expectations (or demands) for tribes to give scientists more genomes; one doesn't build trust that way. And yet, Krystal Tsosie, a Navajo (Diné) geneticist and bioethicist, told me that she attended a conference in early 2019 where a researcher asked her, quite seriously, "What's the magic formula for recruiting Native American participants?"* Others have joked to her about their extensive Native DNA biobanks collected in the past, before tribes had put into place restrictions. She sees attitudes like these and many of the examples discussed above as unified by a common theme:

---

* This is not an isolated incident: Other Indigenous geneticists have told me that they have heard similar questions.

the presumption that the research enterprise's outcomes justify the means by which samples are obtained. "There's just this
sense of white ownership of our Indigenous biological samples,"
she told me. "In some of the more recent controversies you see
non-Indigenous researchers enter into Indigenous spaces. They
collect samples and have complete stewardship over them, using
them for whatever they deem appropriate," rather than working
with tribes themselves to determine how they should be used.

Historically, there has been a decided lack of benefits to
Indigenous groups that have participated in genetics studies.
Criticisms of open-access genomic data have cited this—among
other concerns discussed in this chapter—as a reason for the
lack of tribal participation in genomics studies. There have
been number of initiatives, such as Indigenous-led biobanks and
data repositories like the Native BioData Consortium that prioritize active, ongoing consultation with participants, that show
great promise for improving the balance of power and benefits
between researchers and communities (33).

Pathways Forward

Results from genetics research can have powerful consequences
for Indigenous communities, both beneficial and detrimental.
Scientific research that uses human DNA doesn't take place in a
vacuum. Genetics studies can be used to argue against tribal histories, potentially threaten sovereignty, or dispute cultural identities. They can be used in ways that benefit outside researchers
at the expense of tribal members and potentially reveal stigma-
tizing information.

But there are diverse perspectives among Indigenous

communities, and some feel that the sciences of biological anthropology, archaeology, and genetics could be useful tools in understanding history if used carefully and respectfully, ideally by Indigenous geneticists themselves.

Several tribes look to the ongoing discussions in the Navajo Nation to see how they will end up handling the issue of genetics research; they may very well use any policy that the Navajo develop as a template for creating their own (34).

Others look to the positive examples of work in this field as models for the kind of engagement that is needed.

I'm part of a consortium that has proposed ethical recommendations for paleogenomics research with Indigenous ancestors. We recommended that consultation with Indigenous descendant communities on how paleogenomics research is conducted should be done well before the start of a project. How will the paleogenomic research be conducted? Who will benefit from the results, and how? How will the results be presented? In the absence of known descendants, we recommend that scientists should consult with communities on whose lands the remains are found. Such communities often feel strong ties to—and obligation toward—the ancestors buried within their historic territories, and we believe they are the most appropriate stakeholders in any research concerning these deceased people. These recommendations have been expanded and refined over time as more examples of both positive and negative paleogenomics research have emerged (35).

Throughout this book, we have discussed several cases where genetics research has been viewed positively by Native communities: Shuká K̲áa, Upward Sun River, Nuvuk. There are other cases I have not mentioned in this book where Indigenous groups and tribal representatives have worked very productively with archaeologists and paleogenomics researchers to mutually explore their histories using DNA. For example, in 2017 a research team led by Hendrik Poinar published a mitochondrial DNA–based study of ancient populations from present-day Newfoundland and Labrador. This study showed that multiple groups with distinctive mitochondrial lineages—corresponding to the Maritime Archaic, Paleo-Inuit, and Beothuk archaeological traditions—successively occupied the region, beginning around 4,500 years ago. The study was conducted only after extensive, multiyear consultation with the First Nations and Indigenous peoples who live there today (36). When they published the paper, the research team faced a natural next question: Would they see the same patterns when they looked at complete genomes?

However, as researcher Ana Duggan told me, the team chose not to take that obvious next step, at least not right away. They felt it was crucial to instead go back and further discuss the research with linked communities. "In the past ten years (since the beginning of the project) the technology had changed so much," Duggan told me. The massive amount of data that is produced by sequencing complete ancient genomes can provide far more detailed information on population history, with potential consequences for affiliated communities that may not have been originally anticipated at the project's start. Furthermore, there had been leadership turnovers, and possibly evolving perspectives toward genetics among participating communities. Continuing on with the research, even though technically allowable under the original agreements, "didn't really feel like the right thing to us," Duggan told me. "It felt like something that needed more consultation. We took a step back in part because of the changing environment (political, social, genomic). We're very proud of the work that we did—which we felt was done well— and we would like to continue that trajectory."

This work from the Poinar lab group, as well as many other examples of positive research outcomes have several things in common: a respect for tribal sovereignty and a prioritization of the community's wishes over the scientists' own research agendas, ongoing participation (to various extents) of community members in aspects of the research, and time taken to build relationships rather than conducting "helicopter science." In each case, the knowledge of the past that has been learned from these respectful, reciprocal scientific partnerships has been extraordinary. We must go in this direction.

The path forward in paleogenomics within the Americas will be neither quick nor easy. "We are at the mercy of centuries-old relationships, and it is our duty to recognize and disrupt those harmful legacies," Diné (Navajo) geneticist Justin Lund said at a talk for the American Association of Physical Anthropologists in 2021. Likening the disciplines of genetics and archaeology to houses with crumbling foundations, he advocated for foundational repair by mending relationships with Indigenous groups. "Creating bad relationships took generations, and mending those relationships will also take generations...so plan accordingly! If your work starts and stops in the lab, you're doing it wrong."

Epilogue

Why you so obsessed with me?" an Indigenous anthropologist recently asked me, jokingly channeling Mariah Carey to illustrate the bafflement with which some Indigenous people regard non-Native scholars' obsession with their origins. She meant it humorously but also wanted to convey to me a serious point about the history of our discipline.

Archaeologists, geneticists, and other scholars of the past seek to understand the history and origins of populations across the globe, to answer the question: What does it means to be human, in all our myriad manifestations across time and space? Or, to respond to Mariah Carey with a question posed by the Talking Heads: "How did I get here?"

The ancient history of the Americas holds a special fascination for many of us. This fascination is sparked by the wonder that causes your breath to catch when you see the 10,000-year-old handprints at the Cueva de las Manos site in Patagonia, when you view the exquisitely made 2,000-year-old duck decoys from Lovelock Cave in Nevada, or walk among the silent 700-year-old mounds of Cahokia in Illinois. Who were the people who made these things? What were their lives like?

This is the curiosity that animates me when I'm driving across the plains of Kansas on a consultation trip, hunched over my keyboard and struggling to get the words just right in a grant application, nervously rehearsing my presentation for a tribal council, or bent over a benchtop in a laboratory moving tiny drops of liquid between tubes. I—and I know this is true of my colleagues as well—am profoundly grateful that this is what I get to do for a living.

Technological advancements in archaeology and genetics are allowing us even more insights into how some of humanity's greatest traits—curiosity and ingenuity—led to a people migrating, surviving, and adapting to unknown lands during one of most turbulent episodes in environmental history.

The movement out of Beringia seems as clear to us today as the movement out of Africa, the peopling of Australia, or the multiple migrations of people into Europe. But when we look back through the millennia of human history with knowledge that these journeys happened, we can't help but be influenced by bias: It had to happen this way because this is how it happened. We draw great arrows on maps, confidently pointing southward; we only need to collect some more genomes and do some more analyses in order to fill in the rest of the details.

But this perspective is a foolish one. It erases the reality of the actual history. History is a far messier process, a complicated network of individual choices driven by both the desire for exploration and the necessities of survival. Only from the great distance afforded to us by 13,000 years or more of elapsed time can we see an entire movement of peoples as an arrow on a map.

For many of the descendants of these ancient travelers, the detached assuredness we (non-Native) researchers try to maintain is precisely the problem. The disconnected view of ancient human remains as simply part of natural history, like the fossils of extinct trilobites—the legacy of Thomas Jefferson and the scientists that followed him—underpins the history of genetics and anthropology.

But when scientists turn our lens onto our own history, we are forced to examine ugly things: science used to justify racism, insensitivity in the pursuit of high-impact publications, and atrocities committed in the name of research.

We who work in this field cannot erase our past mistakes—and many of us, myself included, have done research in the past using approaches that we now recognize as wrong. We must acknowledge this, as well as the fact that we have benefited from an unjust system. Only then can we conscientiously address our practices so that our quest to understand ancient humanity helps us maintain our own. In doing so, we must be cognizant of what we are asking for from Indigenous peoples when we design our research.

We are asking for their DNA, which is often considered sacred.

We must consciously ask ourselves, *Am I treating the DNA as sacred?*

We are asking for trust from people whose trust has been violated all too often by our predecessors (and colleagues). We must consciously ask ourselves, *What assurances can I give them that I will uphold this trust now?*

We are asking to destroy small portions of the remains of their ancestors (1).

We must consciously ask ourselves, *How will I ensure that this work is respectful?*

That *I* am respectful?

Whose interests does my research serve?

What benefits will I offer in return?

It is only after we truly grapple with the legacy of our field, interrogate our motives, and deliberately approach our research with intentional respect for the human stories within it that we will able to see the final pieces of the puzzle of the last great step in humanity's journey across the globe.

With all we have learned in recent years in this field, it's easy to lose sight of the forest for the trees. I can't end this book with a simple story, however, and say that it is the final history. Which model for the peopling of the Americas you find most persuasive will depend on how you weight and interpret currently available evidence. Most scholars agree that the ancestors of the First Peoples came from Upper Paleolithic populations in Siberia and East Asia. The precise whereabouts of these ancestors during the LGM—whether in Beringia, eastern Eurasia, the Siberian Arctic Zone, or even in North America south of the Ice Wall—are currently a matter of ongoing research and debate. Ancient DNA shows us that during the LGM, these ancestors remained isolated from other groups in Eurasia and split into several groups, some of which gave rise to the First Peoples south of the Ice Wall. Other groups—like the Ancient Beringians—may have left no contemporary descendants.

There are currently several basic models for how, when, and where people first entered the Americas. The most conservative model resembles a new version of Clovis First: a migration of people belonging to the Diuktai culture in Siberia across the Bering Land Bridge sometime between 16,000 and 14,000 years ago, and south of the Ice Wall—probably down an ice-free corridor—after the LGM. This model is based predominantly on an emphasis of the early Alaskan archaeological record, but does not account for pre-Clovis sites or match the genetic record very well.

The model favored by a small group of archaeologists on the

other end of the spectrum is one in which people came to the Americas very early—some have proposed as early as 130,000 years ago. I (and most archaeologists) do not find this model very convincing, as it has little archaeological support and is completely at odds with what genetics shows us.

The third model—one that I think best fits the totality of both the archaeological and genetic data—posits an entry into the Americas sometime after the first traces of people in Siberia at sites such as Yana around 30,000 years ago. Exactly how early one thinks this may have happened differs depending on which pre-Clovis sites are accepted; the majority of scholars agree that people were present in the Americas by at least 14,000 years ago. Some favor 18,000 to 15,000 years ago to account for the majority of pre-Clovis sites and genetic evidence, and still others argue for a pre-LGM peopling based on archaeological evidence from sites like White Sands Locality 2, and genetic evidence from Population Y. First Peoples most likely traveled by boat along the west coast of North America, reaching South America fairly rapidly.

This is a story without an ending, because as I write this sentence, the genetic story of the Americas is still unfolding. Research that I described in chapter 5 will almost certainly add new details to this story. A few days ago, I was on a conference call with a representative of another community interested in genetics studies. This is another region from which we have very little genetic information—ancient or contemporary—and it's likely that the genomes I might sequence will add to, or change the genetic story even more.

I'm just one obscure researcher from a small lab. Other paleogeneticists, more famous and with vastly more resources than I have, are conducting research constantly, adding large datasets to the literature multiple times a year. "We publish so many papers that we barely even notice anymore when someone in the lab has their paper accepted," one person told me of working in those assembly-line laboratories. The insights coming from these powerhouses are changing our understanding of history so quickly that by the time you read this final chapter, you will almost certainly be able to look back and see out-of-date information.

But this is true of almost every science book, and providing the final, complete story of the peopling of the Americas is not my intention. It's hubris to think that I could do that. Instead, what I hope you'll take from this book is a framework for understanding future developments in the field and an appreciation of the history and complexity that has brought us to this present moment.

Acknowledgments

Many people contributed their time and expertise to the formation of this book. I am indebted first and foremost to the people who taught me how to think scientifically, how to work in the laboratory, how to synthesize different kinds of data, and the importance of viewing the past through different lenses, including Beth and Rudy Raff, José Bonner, Bill Saxton, Rika Kaestle, Geoff Hayes, Deborah Bolnick, and Joane Nagel. A special thanks to Dennis O'Rourke, my first postdoctoral mentor and now my colleague and chair: any professional success I have is because of your boundless generosity and good-humored leadership.

My colleagues at the University of Kansas Department of Anthropology have encouraged and sustained me throughout the writing process. I particularly want to thank Rolfe Mandel, Ivana Radovanović, Fred Sellet, and John Hoopes, who gave me feedback on various aspects of this work.

Members of the Raff/O'Rourke labs made this work possible. Lauren Norman, Justin Tackney, and Kristie Beaty shared their expertise in genetics and archaeology and have supported me throughout the process in countless ways. Savannah Hay formatted my references. Caroline Kisielinski turned my clumsy Power-Point figures into artwork. Other members of the lab read and commented on drafts; I am grateful to each and every one of them (particularly the graduate students).

My colleagues have been extraordinarily generous with their expertise: influencing me with their academic and/or public scholarship, reading and commenting on drafts, contributing images, chatting with me about their ideas, helping me make my writing more sensitive and accurate, and answering my questions. In particular, I'd like to thank Jim Adovasio, Matt Anderson (Eastern Band of Cherokee Indians descent), Jaime Awe, Alyssa Bader (Tsimshian), Jessi Bardill, Rene Begay (Diné), Joan Burke-Kelly, Joe Brewer, Katrina Claw (Diné), Della Cook, Yan Axel Gómez Coutouly, Carlina de la Cova, Michael Crawford, Lee Dugatkin, Ana Duggan, Roger Echo-Hawk (Pawnee), Ken Feder, Terry Fifield, Keolu Fox (Kānaka Maoli), Agustín Fuentes, Ted Goebel, Lynne Goldstein, Kelly Graf, Jessi Halligan, Éadaoin Harney, Leslea Hlusko, John Hoffecker, Eric Hollinger, Jeffrey Hantman, Fatimah Jackson, Anne Jensen, Norma Johnson (Kenaitze), Brian Kemp, David Kilby, Jessica Kolopenuk (Cree, Peguis First Nation), Clark Larsen, Brad Lepper, Justin Lund (Diné), David Meltzer, Kat Milligan-Myhre (Iñupiaq), Amber Nashoba (Choctaw), Maria Nieves-Colón, Dan Odess, Maria Orive, Hendrik Poinar, Bob Sattler, Richard Scott, Mark Sicoli, Pontus Skoglund, Rick Smith, Paulette Steeves (Cree-Metis), Kurly Tlapoyawa (Chicano/Nawa/Mazewalli), Robert Warrior (Osage), Mike Waters, Corey Welch (Northern Cheyenne), Brian Wygal, Patrick Wyman, and all the members of the Beringian working group.

I owe a very special thanks to Scott Elias, Nanibaa' Garrison (Diné), Ripan Malhi, Savannah Martin (Siletz), Kim TallBear (Sisseton-Wahpeton Oyate), Rick Smith, Krystal Tsosie (Diné), Joe Yracheta (P'urhépecha), and members of the SING consortium who have shaped my thinking and helped me in countless ways. I also want to thank all the representatives of tribes and Indigenous organizations who I consulted with on this book, especially Sherry White and Bonney Hartley from the Stockbridge-Munsee Community Band of Mohican Indians, and all my current research collaborators.

My dear friends Ewan Birney, Aylwyn Scally, Stuart Ritchie, Mike Inouye, and Adam Rutherford have been in near-daily conversation with me about topics in genetics, evolution, and race for the last few years. I have learned so much from all of you. Adam has been a particular source of inspiration and encouragement over the years; I am very grateful for our friendship.

Will Francis at Janklow & Nesbit catalyzed this book and gave wonderful notes on many incredibly bad drafts. Working with him has been an absolute pleasure. I'm so grateful to Sean Desmond for his vision and helping me pull together the story I needed to tell. Rachel Kambury helped me see a way through the chaos of the end stages of this book; her advice to "just breathe!" always came exactly when I needed it. Melanie Gold helped keep me organized and was lovely to work with. Thanks to the entire team at Twelve for their patience with a beginning writer.

While I was writing this book, I was also starting a lab, writing grants, going through the tenure process, raising a baby into a toddler, and (for the last two years) managing through a pandemic. I could not have done any of this without Oliver's daycare and preschool teachers (my gratitude to them is beyond expression) and the help of my family. My father, who answered all my questions as a child (in the pre-Internet era) with "Look it up!" has steadfastly encouraged my scientific career and inspired much in this book. In particular, my narrative about caving and geology in chapter 4 is also a tribute to the many hours of underground adventures I had with him in the Ozark Highlands. My

sister Julie, who is my dearest friend, helped support me throughout the writing process and spent hours giving me feedback on
how to make the language in many chapters to be more vivid
and accessible. This book would not be possible without her.
My mother, my first and greatest science teacher, moved in with
us in order to help raise Oliver. I can't thank her enough, nor
Colin, who selflessly sacrificed his own work on many occasions
in order to give me the space to write. To Oliver: you were born
shortly before I started this book and grew into a preschooler as I
worked on it. Although it will be out of date by then, I hope that
you will someday enjoy reading the work that took up so many
of your mother's weekends. You have profoundly influenced my
writing, and I am proud to share this with you.

Notes

A note about citations: This book is intended for non-specialist audiences. Rather than interrupt the flow of sentences by citing references as frequently as would normally be done in an academic text, I chose in many places to group citations in clusters at the end of sections. I also cited what academics call "secondary literature"—review articles and more generalized books about topics—far more often than I would normally do in an academic paper, which relies upon citations to the "primary literature," or research papers and books. The aim of this practice is to guide a reader to accessible overviews of subjects and to limit the frustrating experience of constantly running into paywalls preventing them from finding more information on a subject.

Introduction

1. The history of the discovery and study of Shuká Káa comes from the following sources: E. James Dixon, Timothy H. Heaton, Craig M. Lee, et al., "Evidence of Maritime Adaptation and Coastal Migration from Southeast Alaska," chap. 29 in *Kennewick Man: The Scientific Investigation of an Ancient American Skeleton*, edited by Douglas Owsley and Richard Jantz, 537–548 (Texas A&M University Press, 2014); Sealaska Heritage Institute, "Kuwoot Yas.Ein: His Spirit Is Looking out from the Cave" (Sealaska Heritage Institute, 2005), https://www.youtube.com/watch?v=HDCS56zZaNo; Andrew Lawler, "A Tale of Two Skeletons," *Science* 330, no. 6001 (2010): 171–172, https://doi.org/10.1126/science.330.6001.171, https://science

.sciencemag.org/content/sci/330/6001/171.full.pdf; University of Colorado, Boulder, "Discovery of Ancient Human Remains Sparks Partnership, Documentary," Regents of the University of Colorado, updated September 19, 2006, https://archive.vn/20121214142535/http://www.colorado.edu/news/r/6116efa31a5d9804322e3408e21e1438.html.

2. Brian M. Kemp, Ripan S. Malhi, John McDonough, et al., "Genetic Analysis of Early Holocene Skeletal Remains from Alaska and Its Implications for the Settlement of the Americas," *American Journal of Physical Anthropology* 132, no. 4 (2007): 605–621, https://doi.org/10.1002/ajpa.20543l; John Lindo, Alessandro Achilli, Ugo A. Perego, et al., "Ancient Individuals from the North American Northwest Coast Reveal 10,000 Years of Regional Genetic Continuity," *Proceedings of the National Academy of Sciences* 114, no. 16 (2017): 4093–4098, https://doi.org/10.1073/pnas.1620410114.

3. Rodrigo De los Santos, Cara Monroe, Rico Worl, et al., "Genetic Diversity and Relationships of Tlingit Moieties," *Human Biology* 91, no. 2 (2020): 95–116, https://doi.org/10.13110/humanbiology.91.2.03.

4. There are some notable exceptions, particularly Charles C. Mann, *1491: New Revelations of the Americas Before Columbus* (Knopf, 2005).

5. Roger C. Echo-Hawk, "Ancient History in the New World: Integrating Oral Traditions and the Archaeological Record in Deep Time," *American Antiquity* 65 no. 2 (2000): 267–290.

6. There are many important critical perspectives of archaeology and evolutionary biology by Indigenous scholars, including Vine Deloria Jr., *Red Earth, White Lies: Native Americans and the Myth of Scientific Fact* (Fulcrum Publishing, 1997); Paulette Steeves, *The Indigenous Paleolithic of the Western Hemisphere* (University of Nebraska Press, 2021).

   While a review of the vast literature on Indigenous traditional origin histories is far outside the scope of this book, I do highly recommend Christopher B. Teuton's *Cherokee Stories of the Turtle Island Liars' Club* (University of North Carolina Press, 2012), and Klara Kelley and Harris Francis's *A Diné History of Navajoland* (University of Arizona Press, 2019) as places to get started. A lovely overview of history for children that deftly weaves together traditional narratives with archaeology is *Turtle Island: The Story of North America's First People* by Eldon YellowHorn and Kathy Lowinger (Annick Press, 2017).

7. I thank geneticist and science writer Adam Rutherford for this helpful analogy.

8. I think this requires a bit more explanation here and in chapter 8. As archaeologist Max Friesen notes, "it is a name given to Inuit by outsiders rather than a self-designation, and it has come to be considered pejorative

in some, though certainly not all, contexts." He recommends instead the use of the terms *Inuit* or *Paleo-Inuit*, and in the same spirit I will be referring to the language group Inuit-Aleut throughout. T. Max Friesen, "On the Naming of Arctic Archaeological Traditions: The Case for Paleo-Inuit," *Arctic* 68, no. 3 (2015): iii–iv. But I want to stress that this preference is by no means universally shared among Indigenous Arctic peoples. As with all names in this book, it is a complicated issue. My approach in day-to-day practice is to ask what term an individual person wishes me to use for them and use it. For this book I have consulted with and followed the recommendations of Indigenous colleagues in genetics and archaeology.

Chapter 1

1. Edwin Hamilton Davis and George Ephraim Squier, *Ancient Monuments of the Mississippi Valley: Comprising the Results of Extensive Original Surveys and Explorations* (Smithsonian Institution, 1848).
2. Bradley T. Lepper and Tod A. Frolking, "Alligator Mound: Geoarchaeological and Iconographical Interpretations of a Late Prehistoric Effigy Mound in Central Ohio, USA," *Cambridge Archaeological Journal* 13, no. 2 (2003): 147–167, https://doi.org/10.1017/S0959774303000106.
3. Bradley T. Lepper, "Archaeology: Serpent Mound Might Depict a Creation Story," Last modified: February 11, 2018, https://www.dispatch.com/news/20180211/archaeology-serpent-mound-might-depict-creation-story; William H. Holmes, "A Sketch of the Great Serpent Mound," *Science* 8, no. 204 (1886): 624–628; Robert V. Fletcher, Terry L. Cameron, Bradley T. Lepper, et al., "Serpent Mound: A Fort Ancient Icon?" *Midcontinental Journal of Archaeology* (1996): 105–143.
4. References for this section: Jason Colavito, *The Mound Builder Myth: Fake History and the Hunt for a "Lost White Race"* (University of Oklahoma Press, 2020); Kenneth L. Feder, *Frauds, Myths, and Mysteries: Science and Pseudoscience in Archaeology*, 10th ed. (Oxford University Press, 2020); David H. Thomas, *Skull Wars: Kennewick Man, Archaeology, and the Battle for Native American Identity* (Basic Books, 2000).
5. Brook Wilensky-Lanford, "The Serpent Lesson: Adam and Eve at Home in Ohio," *The Common*, last modified: March 1, 2011, https://www.thecommononline.org/the-serpent-lesson-adam-and-eve-at-home-in-ohio.
6. Thomas, *Skull Wars*.
7. José de Acosta, *Historia Natural y Moral de Las Indias*[...] (Seville, 1590).
8. This and other details about Jefferson's response to de Buffon can be

found in Lee Dugatkin's excellent book *Mr. Jefferson and the Giant Moose: Natural History in Early America* (University of Chicago Press, 2009).

9. Thomas, *Skull Wars*.

10. Thomas Jefferson, *Notes on the State of Virginia* (New York: M. L. & W. A. Davis, 1801).

11. Sources for this section include Debra L. Gold, *The Bioarchaeology of Virginia Burial Mounds* (University of Alabama Press, 2004); Jeffrey L. Hantman, "Monacan Archaeology of the Virginia Interior, A.D. 1400–1700," pp. 107–124 in *Societies in Eclipse: Archaeology of the Eastern Woodlands Indians, A.D. 1400–1700*, edited by David S. Brose, C. Wes Cowan, and Robert C. Mainfort Jr. (University of Alabama Press, 2001); Jeffrey L. Hantman, *Monacan Millennium: A Collaborative Archaeology and History of a Virginia Indian People* (University of Virginia Press, 2018); Jeffrey L. Hantman, "Between Powhatan and Quirank: Reconstructing Monacan Culture and History in the Context of Jamestown," *American Anthropologist* 92, no. 3 (1990): 676–690, https://doi.org/10.1525/aa.1990.92.3.02a00080; Philip Barbour, ed., *The Complete Works of Captain John Smith (1580–1631)*, 3 vols. (University of North Carolina Press, 1986). For more on Monacan history and the mounds, see Monacan Indian Nation, "Our History," https://www.monacannation.com/our-history.html.

12. Thomas Jefferson, *Notes on the State of Virginia*, 1832 printing, pp. 103–104.

13. For a thorough treatment of the Mound Builder hypothesis, I recommend Colavito, *The Mound Builder Myth*.

14. Journal of the Senate of the United States of America being the First Session of the Twenty-First Congress Begun and Held at the City of Washington. S. Rep. (Washington, DC: Duff Green, 1829).

15. Thomas, *Skull Wars*.

16. Samuel J. Redman, *Bone Rooms* (Harvard University Press, 2016), 72–73.

17. George G. Heye and George H. Pepper, "Exploration of a Munsee Cemetery Near Montague, New Jersey," in *Contributions from the Museum of the American Indian Volume II (1)* (Heye Foundation, 1915–1916), p. 3.

18. The court's decision was reprinted in the April-June 1915 issue of *American Anthropologist* (p. 415) with the comment "The reversal by the higher court of the conviction of George G. Heye for the removal of skeletons from an old Indian burial ground in New Jersey is a matter of interest to all persons engaged in archaeological work."

19. See https://naturalhistory.si.edu/education/teaching-resources/social-studies/forensic-anthropology.

20. But see Elizabeth A. DiGangi and Jonathan D. Bethard, "Uncloaking a Lost Cause: Decolonizing Ancestry Estimation in the United States,"

American Journal of Physical Anthropology 175 (2021): 422–436, https://doi
.org/10.1002/ajpa.

21. For a thorough discussion of this issue by a Black biological anthropologist, see Carlina de la Colva, "Marginalized Bodies and the Construction
of the Robert J. Terry Anatomical Skeletal Collection: A Promised Land
Lost," in *Bioarchaeology of Marginalized Peoples,* edited by Madeleine Mant
and Alyson Holland (Cambridge, MA: Elsevier, 2019). Also see "Haunted
by My Teaching Skeleton," by Asian American primatologist and human
biologist Michelle Rodrigues, https://www.sapiens.org/archaeology
/where-do-teaching-skeletons-come-from/.

22. Review of Elizabeth Weiss and James W. Springer's book *Repatriation and
Erasing the Past,* https://journals.kent.ac.uk/index.php/transmotion/arti
cle/view/993/1919. Smiles also discusses issues regarding Indigenous
sovereignty over their own bodies in a related context—state-mandated
autopsies—in a recent article, "'... to the Grave'—Autopsy, Settler Structures, and Indigenous Counter-Conduct," *Geoforum* 91 (2018): 141–150.

23. Aleš Hrdlička, "The Genesis of the American Indian," in *Proceedings of the
Nineteenth International Congress of Americanists* (International Congress of
Americanists, 1917).

24. Aleš Hrdlička, *Physical Anthropology of the Lenape or Delawares, and of the Eastern Indians in General* (Government Printing Office, 1916), doi: https://doi
.org/10.5479/sil.451251.39088016090649.

25. It is arguably still entrenched today. The history of race and anthropology is a vast subject, one which I can't do justice to in the short summary
offered here. This survey focuses on race and the history of physical
anthropology in the United States as it pertains to Native American origins, using the following sources: Michael L. Blakey, "Intrinsic Social and
Political Bias in the History of American Physical Anthropology: With
Special Reference to the Work of Aleš Hrdlička," *Critique of Anthropology* 7,
no. 2 (1987): 7–35, https://doi.org/10.1177/0308275X8700700203; Rachel
Caspari, "Race, Then, and Now: 1918 Revisited," *American Journal of Physical Anthropology* 165, no. 4 (2018): 924–938; Rachel Caspari, "Deconstruct-
ing Race: Race, Racial Thinking, and Geographic Variation, and the
Implications for Biological Anthropology," pp. 104–122 in *A Companion
to Biological Anthropology,* edited by Clark S. Larsen (John Wiley & Sons,
2010); Michael A. Little and Robert W. Sussman, "History of Biological
Anthropology"; Joseph F. Powell, *The First Americans: Race, Evolution, and
the Origin of Native Americans* (Cambridge University Press, 2005); Adam
Rutherford, *How to Argue with a Racist: History, Science, Race, and Reality*
(Weidenfeld & Nicolson, 2020); Thomas, *Skull Wars*; Michael H. Crawford,

The Origins of Native Americans: Evidence from Anthropological Genetics (Cambridge University Press, 1998); John S. Michael, "A New Look at Morton's Craniological Research," *Current Anthropology* 29, no. 2 (1988): 349–354, https://doi.org/10.1086/203646; Jonathan Marks, *Tales of the Ex-Apes: How We Think About Human Evolution* (University of California Press, 2015); Jonathan Marks, *Human Biodiversity: Genes, Race, and History* (Aldine de Gruyter, 1995); Sheela Athreya and Rebecca Rogers Ackermann, "Colonialism and Narratives of Human Origins in Asia and Africa" in *Interrogating Human Origins: Decolonisation and the Deep Past* (Routledge, 2019); *A History of American Physical Anthropology 1930–1980*, edited by Frank Spencer (Academic Press, 1982), particularly chapter 1 (by C. Loring Brace), chapter 11 (by Albert B. Harper and William S. Laughlin), and chapter 12 (by George J. Armelagos, David S. Carlson, and Dennis P. Van Gerven); Agustín Fuentes, *Race, Monogamy and Other Lies They Told You: Busting Myths About Human Nature* (University of California Press, 2012); Sherwood Washburn, "The New Physical Anthropology," *Transactions of the New York Academy of Science* 13 (1951): 298–304; *Histories of American Physical Anthropology in the Twentieth Century,* edited by Michael A. Little and Kenneth A. R. Kennedy (Lexington Books, 2010).

26. Morton believed that because Black Africans had such small brains, they were natural slaves. His measurements have been heavily critiqued as systematically biased, most famously by the late paleontologist Steven J. Gould in his 1981 book *The Mismeasure of Man* (W.W. Norton & Company, 1981). Others have critiqued Gould's critique. But most experts I've talked to on the subject who have actually worked with the Morton collection believe that his analyses were indeed significantly biased to inflate the average size of European crania and depress the calculated average size for all other races. In "The Fault in His Seeds: Lost Notes to the Case of Bias in Samuel George Morton's Cranial Race Science" (*PLOS Biology* 16, no. 10 (2018): e2007008, https://doi.org/10.1371/journal.pbio.2007008), Paul Wolff Mitchell argues that Gould's accusation of bias in Morton's data was wrong, but that Gould was correct about Morton's perspectives, intentions, and use of his data to promote racist ends.

27. Samuel George Morton, *Crania Americana, or, A Comparative View of the Skulls of Various Aboriginal Nations of North and South America* (J. Dobson, 1839).

28. In this book I focus my very brief survey on the history of physical anthropology in the United States, but the reader should understand that there are strong traditions of physical anthropology in other countries across the world.

The terms *physical anthropology* and *biological anthropology* are often used synonymously. However, the term *biological anthropology* is increasingly favored by researchers in the United States to reflect a break from its more typological past and a focus on, as Agustín Fuentes notes in a 2021 article, "the human experience in the broader ecological, evolutionary, and phylogenetic context" (see "Biological Anthropology's Critical Engagement with Genomic Evolution, Race/Racism, and Ourselves: Opportunities and Challenges to Making a Difference in the Academy and the World," *American Journal of Physical Anthropology* 175, no. 2 (2021): 326–338). Throughout this book, I will refer to the field as physical anthropology when discussing its early history, and biological anthropology for more recent events, reflecting the discipline's evolution.

29. Blakey, "Intrinsic Social and Political Bias"; Michael L. Blakey, "Understanding Racism in Physical (Biological) Anthropology," *American Journal of Physical Anthropology*, https://doi.org/10.1002/ajpa.24208.

30. Franz Boas, "Changes in the Bodily Form of Descendants of Immigrants," *American Anthropologist* 14, no. 3 (1912): 530–562, https://doi.org/10.1525/aa.1912.14.3.02a00080.

31. Heather Pringle, *The Master Plan: Himmler's Scholars and the Holocaust* (Hyperion Books, 2006); B. Müller-Hill, *Murderous Science: Elimination by Scientific Selection of Jews, Gypsies, and Others, Germany 1933–1945* (Oxford University Press, 1988).

32. There were some notable exceptions, foremost among them Carlton Coon, who continued to classify races and subraces in his books *The Origin of Races* (Alfred A. Knopf, 1962) and (with Edward E. Hunt) *The Living Races of Man* (Alfred A. Knopf, 1965). His work was widely repudiated by his peers.

33. Cobb is reputed to have Native American ancestry through his mother, but according to Dr. Fatimah Jackson, the curator of the Cobb collection at Howard University, this has not been documented. Many Black North American families have oral traditions of Native ancestry. This complex subject is outside the scope of this book, but I refer the interested reader to Jackson's article "What Is Wrong with African North American Admixture Studies? Addressing the Questionable Paucity of Amerindian Admixture in African North American Genetic Lineages," *Journal of Genetics and Cell Biology* 4, no. 1 (2020): 228–232.

References for this section include: W. Montague Cobb, "Race and Runners," *Journal of Health and Physical Education* 7, no. 1 (1936): 3–56; W. Montague Cobb, "Physical Anthropology of the American Negro," *American Journal of Physical Anthropology* 29, no. 2 (1942): 113–223; Rachel J. Watkins, "'[This] System Was Not Made for [You]': A Case for

Decolonial Scientia," *American Journal of Physical Anthropology*, https://doi
.org/10.1002/ajpa.24199.

34. I find some reason to be optimistic in the evolution of my professional
organization. The American Association of Physical Anthropology has
embraced its responsibility to atone for the harms that it has caused,
first with a new statement on race and racism (co-authored by Rebecca
Ackermann, Sheela Athreya, Deborah Bolnick, Agustín Fuentes, Tina
Lasisi, Sang-Hee Lee, Shay-Akil McLean, and Robin Nelson) and then
in solidarity with civil rights protestors an open letter that gave very specific recommendations to its members to "counter the impact óf harmful
work done by our professional predecessors and to call out scientific racism today." The entire statement can be read here: https://physanth.org
/about/position-statements/open-letter-our-community-response-police
-brutality-against-african-americans-and-call-antiracist-action. As a symbolic action of this break with past traditions and practices, the AAPA has
just changed its name to the American Association of Biological Anthropologists. In addition, there has been a groundswell of outstanding scholarship within the discipline that deserves attention and credit. In 2019,
the AAA Vital Topics forum published a series of articles "to explore the
ways that scientists from diverse backgrounds are producing new, exciting, and essential kinds of knowledge about humans and nonhumans; the
connections between bodies, biology and culture; and the politics and
practices of science" (Deborah A. Bolnick, Rick W. A. Smith, and Agustín
Fuentes, "How Academic Diversity Is Transforming Scientific Knowledge
in Biological Anthropology," *American Anthropologist* 121 (2019): 464,
http://doi.org/10.1111/aman.13212). In June 2021, the *American Journal of
Physical Anthropology* published a special issue on race, organized by me
and Connie Mulligan, with contributions from outstanding scholars in
the discipline that address everything from how white supremacists use
the knowledge produced by biological anthropologists to how forensic anthropology's typological approaches reinforce racial categories.
Although great change is still needed within the AABA, many within the
organization (particularly some early-career scholars) take this responsibility very seriously and are making a great impact.

35. See Ian Mathieson and Aylwyn Scally, "What Is Ancestry?" *PLOS Genetics*
(2020), https://doi.org/10.1371/journal.pgen.1008624; Nick Patterson,
Priya Moorjani, Yontao Luo, et al., "Ancient Admixture in Human History," *Genetics* 192 (2012): 1065–1093, https://doi.org/10.1534/genetics
.112.145037; and Graham Coop's excellent blog post, "How Many Genetic

Ancestors Do I Have?," https://gcbias.org/2013/11/11/how-does-your-num ber-of-genetic-ancestors-grow-back-over-time/.

36. For a debunking of the biological race concept, see references in (25) above as well as the American Society of Human Genetics's statement "ASHG Denounces Attempts to Link Genetics and Racial Supremacy," *American Journal of Human Genetics* 103, no. 5 (2018): 636, doi: 10.1016/j. ajhg.2018.10.011; Lynn B. Jorde and Stephen P. Wooding, "Genetic Varia- tion, Classification and 'Race'," *Nature Genetics* 36 (2004), S28–S33; Gar- rett Hellenthal, George B. J. Busby, Gavin Band, et al., "A Genetic Atlas of Human Admixture History," *Science* 343, no. 6172 (2014): 747–751, doi: 10.1126/science.1243518; and a blog post written by Ewan Birney, me, Adam Rutherford, and Aylwyn Scally entitled "Race, Genetics and Pseu- doscience: An Explainer" (http://ewanbirney.com/2019/10/race-genetics -and-pseudoscience-an-explainer.html).

37. See "The Biology of Racism," a published discussion by Leith Mullings, Jada Benn Torres, Agustín Fuentes, et al., *American Anthropologist* (2021), doi: 10.1111/aman.13630.

38. Redman, *Bone Rooms.*

39. Hrdlička, *Physical Anthropology of the Lenape or Delawares*, vol. 3. More than 100 years later, I'm struck by how much Hrdlička got wrong about these individuals. Contrary to Hrdlička's assertion, disease and malnutrition have left extensive evidence on the bones of these ancestors. It's actually quite hard to read about these pathologies, knowing how much the peo- ple who carried them would have suffered, and given that these condi- tions were not present to nearly the same extent on remains from this region prior to European contact, it's hard not to attribute them to the effects of colonization. Techniques for studying the human skeleton and identifying the effects of disease and trauma began in the nineteenth cen- tury but became far more sophisticated and standardized in the second half of the twentieth century. For introductions into the history of this field, see: Jane Buikstra, "Paleopathology: A Contemporary Perspective," in *A Companion to Biological Anthropology*, edited by Clark S. Larsen (John Wiley & Sons, 2010), pp. 395–311; Della C. Cook and Mary Lucas Powell, "The Evolution of American Paleopathology," in *Bioarchaeology: The Con- textual Analysis of Human Remains*, edited by Jane E. Buikstra and L. E. Beck (Academic Press, 2006), pp. 281–322; Anne L. Grauer, "A Century of Paleopathology," *American Journal of Physical Anthropology* 165 (2018): 904, DOI: 10.1002/ajpa.23366.

Today the remains of the Minisink Lenape have been reburied with their funerary items in a safe spot chosen by their descendants. The return of their ancestors' remains to the Stockbridge-Munsee, Delaware Nation, and Delaware Tribe was the result of years of dialogue between the tribes and government officials. Brice Obermeyer, "Repatriations and Culture Camp Planned for 2015," last modified January 25, 2015, Delawaretribe .org/blog/2015/01/25/repatriations-and-culture-camp.

For a thorough overview of the archaeology of these ancestors, see Herbert C. Kraft's *The Lenape-Delaware Indian Heritage: 10,000 B.C.–A.D. 2000*, chapter 6. This book also provides an excellent starting point for understanding the history of present-day descendants. I also encourage interested readers to visit tribal websites for more information.

40. C. G. Turner II, "Dental Evidence for the Peopling of the Americas," pp. 147–157 in *Early Man in the New World*, edited by Richard Shutler Jr. (Sage, 1983); Joseph F. Powell, *The First Americans: Race, Evolution and the Origin of Native Americans* (Cambridge University Press, 2005).

41. Hrdlička, "The Genesis of the American Indian."

42. D. H. O'Rourke, "Blood Groups, Immunoglobulins, and Genetic Variation," in *Handbook of North American Indians*, edited by Douglas H. Ubelaker (Smithsonian Institution, 2006), pp. 762–776.

43. E. J. E. Szathmáry, "Genetics of Aboriginal North Americans," *Evolutionary Anthropology* 1 (1993): 202–220.

44. The literature on American mitochondrial haplogroups is too large to include comprehensively here. For overviews of these haplogroups, see: Rafael Bisso-Machado and Nelson J. R. Fagundes, "Uniparental Genetic Markers in Native Americans: A Summary of All Available Data From Ancient and Contemporary Populations," *American Journal of Physical Anthropology* (2021), https://doi.org/10.1002/ajpa.24357; Dennis H. O'Rourke and Jennifer A. Raff, "The Human Genetic History of the Americas: The Final Frontier," *Current Biology* 20, no. 4 (2010): R202–207, https://doi.org/10.1016/j.cub.2009.11.051. Additional references for this section include the following articles (and references therein): Ugo A. Perego, Alessandro Achilli, Norman Angerhofer, et al., "Distinctive Paleo-Indian Migration Routes from Beringia Marked by Two Rare mtDNA Haplogroups," *Current Biology* 19, no. 1 (2009): 1–8; Bastien Llamas, Lars Fehren-Schmitz, Guido Valverde, et al., "Ancient Mitochondrial DNA Provides High-Resolution Time Scale of the Peopling of the Americas," *Science Advances* 2, no. 4, (2016), e1501385, DOI: 10.1126/sciadv.1501385.

45. For a very thorough debunking of Graham Hancock's book, see the November 2019 (vol. 19, no. 5) issue of the Society for American Archaeology's *Archaeological Record*, especially Jason Colavito's article

"Whitewashing American Prehistory" (*SAA Archaeological Record* 19, no. 5 [2019]: 17–20).

46. The two most prominent proponents of the Solutrean hypothesis are the late Dennis Stanford of the Smithsonian Museum and Bruce Bradley, emeritus professor at the University of Exeter. They published the case for this theory in their 2012 book *Across Atlantic Ice: The Origin of America's Clovis Culture* (University of California Press, 2012).

47. There are many archaeological rebuttals to this model, but here are two excellent sources: Lawrence Guy, David J. Meltzer, and Ted Goebel, "Ice Age Atlantis? Exploring the Solutrean-Clovis 'Connection,'" *World Archaeology* 37, no. 4 (2005): 507–532, https://doi.org/10.1080/00438240500395797; Lawrence G. Straus, "Solutrean Settlement of North America? A Review of Reality," *American Antiquity* 65, no. 2 (2017): 219–226, doi:10.2307/2694056.

48. Metin I. Eren, Robert J. Patten, Michael J. O'Brien, David J. Meltzer, "Refuting the Technological Cornerstone of the Ice-Age Atlantic Crossing Hypothesis," *Journal of Archaeological Science* 40, no. 7 (2013): 2934–2941.

49. Eske Willerslev and David J. Meltzer, "Peopling of the Americas as Inferred from Ancient Genomes," *Nature* 594, no. 7863 (2021): 356–364, doi: 10.1038/s41586-021-03499-y.

50. For more on the L'Anse Aux Meadows site, read Paul M. Ledger, Linus Girdland-Flink, and Véronique Forbes, "New Horizons at L'Anse Aux Meadows," *Proceedings of the National Academy of Sciences* 116, no. 31 (2019): 15341–15343, https://doi.org/10.1073/pnas.1907986116.

51. The paper that posited the hypothesis that mitochondrial haplogroup X is genetic evidence of a Solutrean migration, and its rebuttal: Stephen Oppenheimer, Bruce Bradley, and Dennis Stanford, "Solutrean Hypothesis: Genetics, the Mammoth in the Room," *World Archaeology* 46, no. 5 (2014): 752–774, https://doi.org/10.1080/00438243.2014.966273; Jennifer A. Raff and Deborah A. Bolnick, "Does Mitochondrial Haplogroup X Indicate Ancient Trans-Atlantic Migration to the Americas? A Critical Re-evaluation," *PaleoAmerica* 1, no. 4 (2015): 297–304, https://doi.org/10.1179/2055556315Z.00000000040.

Chapter 2

1. John F. Hoffecker, *Modern Humans: Their African Origin and Global Dispersal* (Columbia University Press, 2017); Shawn J. Marshall, Thomas S. James, Garry K.C. Clarke, "North American Ice Sheet Reconstructions at the Last Glacial Maximum," *Quaternary Science Reviews* 21 (2002): 175–192.

2. Jenni Lanham, "Folsom, NM Storm and Flooding, Aug 1908—Telephone

Operator a Hero," accessed May 17, 2020, http://www.gendisasters.com
/new-mexico/9309/folsom-nm-storm-flooding-aug-1908-telephone
-operator-hero. See also David Meltzer's *First Peoples in a New World: Colonizing Ice Age America* (University of California Press, 2010).

3. Yet despite his position of authority, he was reportedly referred to
"affectionately" with a racial slur preceding his first name. "Discovered
Folsom Man: A Nomadic Hunter Who Roamed New Mexico More than
10,000 Years Ago," accessed May 17, 2020, http://www.folsomvillage.com
/folsommuseum/georgemcjunkin.html. For more on George McJunkin,
see J. M. Adovasio and Jake Page, *The First Americans: In Pursuit of Archaeology's Greatest Mystery* (Random House, 2002); Christina Proenza-Coles's *American Founders: How People of African Descent Established Freedom in the New World*
(NewSouth Books: 2019), p. 230; and Meltzer, *First Peoples in a New World*.

4. David Meltzer's book *The Great Paleolithic War: How Science Forged an Understanding of America's Ice Age Past* (University of Chicago Press, 2015) is
essential reading about this history; the quote is from page 11.

5. R. E. Taylor, "The Beginnings of Radiocarbon Dating in *American Antiquity*: A Historical Perspective," *American Antiquity* 50, no. 2 (1985): 309–325,
https://doi.org/10.2307/280489.

6. References for this history include William H. Holmes, "The Antiquity
Phantom in American Archeology," *Science* 62, no. 1603 (1925): 256–258;
David J. Meltzer, *First Peoples in a New World*; Aleš Hrdlička, "The Com-
ing of Man from Asia in the Light of Recent Discoveries," *Proceedings of the
American Philosophical Society* 71, no. 6 (1932): 393–402; David J. Meltzer,
"On 'Paradigms' and 'Paradigm Bias' in Controversies over Human Antiq-
uity in America," *The First Americans: Search and Research* (1991): 13–49;
Michael R. Waters, *Principles of Geoarchaeology: A North American Perspective*
(University of Arizona Press, 1992); Harold J. Cook, "Glacial Age Man
in New Mexico," *Scientific American* 139, no. 1 (1928): 38–40, https://doi
.org/10.1038/scientificamerican0728-38.

7. I am indebted to my colleague and friend Rolfe Mandel for helping me
with this explanation of the field in which he is an expert. I also relied
upon Michael R. Waters, *Principles of Geoarchaeology*.

8. Harold J. Cook, "Glacial Age Man in New Mexico," *Scientific American* 139,
no. 1 (1928): 38–40.

9. For a more complete discussion of this topic see Meltzer, *First Peoples in a
New World*; Thomas, *Skull Wars*.

10. Meltzer, *The Great Paleolithic War*.

11. C. Vance Haynes, "Fluted Projectile Points: Their Age and Dispersion: Stratigraphically Controlled Radiocarbon Dating Provides New Evidence on

Peopling of the New World," *Science* 145, no. 3639 (1964): 1408–1413; P. S. Martin, "The Discovery of America," *Science* 179, no. 4077 (1973): 969–974.

12. Dates in archaeology are very confusing. If you go back and read the archaeological literature, you will find a bewildering number of dates presented for the same periods. Is Clovis dated to 11,500 years ago? 12,500? 12,700? 13,000? The dates given in a publication depend on what methods were used for dating, whether they were calibrated or in radiocarbon years, what methods were used for calibration, which sites the authors accept as legitimate, and so forth. Rather than confuse you (and myself) by constantly inserting qualifiers every time I mention a date, I'm going to be explicit about which sources I use here containing the most current estimates of dates that I trust. These are derived from two good reviews on the subject: one by geoarchaeologist Michael R. Waters, "Late Pleistocene Exploration and Settlement of the Americas by Modern Humans," *Science* 365, no. 6449 (2019), and the other by geneticist Eske Willerslev and archaeologist David J. Meltzer in "Peopling of the Americas As Inferred from Ancient Genomes." Dates for Clovis come from a 2020 paper by Michael R. Waters, Thomas W. Stafford Jr., and David L. Carlson, "The Age of Clovis—13,050 to 12,750 cal yr B.P."

This means that when I'm talking about certain events in the history of archaeology—like the acceptance of Monte Verde as a pre-Clovis site—I'm going to be using contemporary dates in place of the ones that were originally reported at the time these events took place. For instance, the dates I present here in this Clovis First scenario are not actually the ones my archaeology professor gave us in class when I took it in the early 2000s; they've been updated to reflect the most current dates for Clovis, the opening of the ice-free corridor, and so forth. I hope you excuse this liberty, but otherwise I suspect many readers will become very confused. So if you go back and read the original publications, the dates discussed in those pages may differ significantly from what I'm writing here. This also means that if you're reading this book well after its publication, all the dates I'm giving here might be considered wrong. Even at the time of publication, I expect there will be some archaeologists who disagree with the dates that I'm giving here.

13. The dates for Clovis—as with any other technocomplex in the archaeological record—depend on which sites are determined to *be* Clovis, what radiocarbon ages are considered to be accurate (e.g., free of contamination, in geological context that is undisturbed, from appropriate material), and what calibration methods are used. Choices are made about which sites are included in the Clovis complex, and these choices have

major implications for understanding history. In 2020, Michael R. Waters, Thomas Stafford Jr., and David L. Carlson published a paper entitled "The Age of Clovis—13,050 to 12,750 cal yr B.P" in which they reported new radiocarbon dates for some key Clovis sites and produced a new chronology for Clovis based on only those sites that had diagnostic Clovis points present. Their new age for Clovis, 13,050–12,750 cal years BP (or "years ago") is what I use throughout this book. Not all archaeologists are likely to agree on these dates.

14. James M. Adovasio and Ronald C. Carlisle, "The Meadowcroft Rockshelter," *Science* 239, no. 4841 (1988): 713–714; J. M. Adovasio, J. Donahue, and R. Stuckenrath, "The Meadowcroft Rockshelter Radiocarbon Chronology 1975–1990," *American Antiquity* 55, no. 2 (1990): 348–354, doi:10.2307/281652; J. M. Adovasio, J. D. Gunn, J. Donahue, et al., "Meadowcroft Rockshelter, 1977: An Overview," *American Antiquity* 43, no. 4 (1978): 632–651, doi:10.2307/279496; J. M. Adovasio, J. D. Gunn, J. Donahue, et al., "Yes Virginia, It Really Is That Old: A Reply to Haynes and Mead," *American Antiquity* 45, no. 3 (1980): 588–595, doi:10.2307/279879; C. Vance Haynes, "Paleoindian Charcoal from Meadowcroft Rockshelter: Is Contamination a Problem?" *American Antiquity* 45, no. 3 (1980): 582–587, doi:10.2307/279878; Kenneth B. Tankersley and Cheryl Ann Munson, "Comments on the Meadowcroft Rockshelter Radiocarbon Chronology and the Recognition of Coal Contaminants," *American Antiquity* 57, no. 2 (1992): 321–326, https://doi.org/10.2307/280736; James Adovasio and Jake Page, *The First Americans: In Pursuit of Archaeology's Greatest Mystery* (Random House, 2002).

15. Joseph H. Greenberg, Christy G. Turner, Stephen L. Zegura, et al., "The Settlement of the Americas: A Comparison of the Linguistic, Dental, and Genetic Evidence [and Comments and Reply]," *Current Anthropology* 27, no. 5 (1986): 477–497, https://doi.org/10.1086/203472; Deborah A. Bolnick, Beth A. (Shultz) Shook, Lyle Campbell, et al., "Problematic Use of Greenberg's Linguistic Classification of the Americas in Studies of Native American Genetic Variation," *American Journal of Human Genetics* 74, no. 3 (2004): 519–523; Johanna Nichols, "Linguistic Diversity and the First Settlement of the New World," *Language* 66, no. 3 (1990): 475–521; David Reich, Nick Patterson, Desmond Campbell, et al., "Reconstructing Native American Population History," *Nature* 488 (2012): 370–374, https://doi.org/10.1038/nature11258.

16. Tom D. Dillehay and Michael B. Collins, "Early Cultural Evidence from Monte Verde in Chile," *Nature* 332, no. 6160 (1988): 150–152, https://doi.org/10.1038/332150a0; T. D. Dillehay, C. Ramirez, M. Pino, et al., "Monte Verde: Seaweed, Food, Medicine, and the Peopling of South

America," *Science* 320, no. 5877 (2008): 784–786, https://doi.org/10.1126/science.1156533; Thomas D. Dillehay, *The Settlement of the Americas: A New Prehistory* (Basic Books, 2000).

17. This historical event in archaeology is described vividly by David Meltzer in two publications: David J. Meltzer, Donald K. Grayson, Gerardo Ardila, et al., "On the Pleistocene Antiquity of Monte Verde, Southern Chile," *American Antiquity* 62, no. 4 (1997): 659–663, https://doi.org/10.2307/281884; and Meltzer, *First Peoples in a New World*. Another perspective on this event is given by J. M. Adovasio and Jake Page in *The First Americans*. The unanimity of the archaeologists' agreement on the antiquity of Monte Verde was later undermined by the retraction of Vance Haynes, who fixated on the six stone tools as the "only unequivocal artifacts" and thought that was insufficient evidence to support a pre-Clovis date. As Meltzer notes in *First Peoples in a New World*, that ignores all the organic artifacts at the site, and "Haynes's second thoughts have not reversed the archaeological tide of opinion on Monte Verde" (p. 128). Some archaeologists, most notably Stuart Fiedel, continue to disagree with the prevailing view that Monte Verde is pre-Clovis: Stuart J. Fiedel, "Is That All There Is? The Weak Case for Pre-Clovis Occupation of Eastern North America," in *The Eastern Fluted Point Tradition*, edited by Joseph A. M. Gingerich (University of Utah Press, 2013), pp. 333–354.

18. M. Thomas P. Gilbert, Dennis L. Jenkins, Anders Götherstrom, et al., "DNA from Pre-Clovis Human Coprolites in Oregon, North America," *Science* 320, no. 5877 (2008): 786, https://doi.org/10.1126/science.1154116; S. David Webb, *First Floridians and Last Mastodons: The Page-Ladson Site in the Aucilla River* (Springer Netherlands, 2006); Thomas A. Jennings and Michael R. Waters, "Pre-Clovis Lithic Technology at the Debra L. Friedkin Site, Texas: Comparisons to Clovis through Site-Level Behavior, Technological Trait-List, and Cladistic Analyses," *American Antiquity* 79, no. 1 (2014): 25–44, https://doi.org/10.7183/0002-7316.79.1.25; Loren G. Davis, David B. Madsen, Lorena Becerra-Valdivia, et al., "Late Upper Paleolithic Occupation at Cooper's Ferry, Idaho, USA, ~16,000 Years Ago," *Science* 365, no. 6456 (2019): 891–897, DOI: 10.1126/science.aax9830; *Taima-Taima: A Late Pleistocene Paleo-Indian Kill Site in Northernmost South America. Final Report of 1976 Excavations*, edited by C. Ochsenius and R. Gruhn (Programa CIPICS, Monografías Científicas, Universidad Francisco de Miranda, 1979).

19. Michael R. Waters, Steven L. Forman, Thomas A. Jennings, et al., "The Buttermilk Creek Complex and the Origins of Clovis at the Debra L. Friedkin Site, Texas," *Science* 331, no. 6024 (2011): 1599–1603.

20. Matthew R. Bennett, David Bustos, Jeffrey S. Pigati, et al., "Evidence of

Humans in North America during the Last Glacial Maximum," *Science* (2021), in press.

21. Peter D. Heintzman, Duane Froese, John W. Ives, et al., "Bison Phylogeography Constrains Dispersal and Viability of the Ice Free Corridor in Western Canada," *Proceedings of the National Academy of Sciences* 113, no. 29 (2016): 8057–8063, https://doi.org/10.1073/pnas.1601077113.

22. Ben A. Potter, James F. Baichtal, Alwynne B. Beaudoin, et al., "Current Evidence Allows Multiple Models for the Peopling of the Americas," *Science Advances* 4, no. 8 (2018): eaat5473, https://doi.org/10.1126/sciadv.aat5473.

23. Heather L. Smith and Ted Goebel, "Origins and Spread of Fluted-Point Technology in the Canadian Ice-Free Corridor and Eastern Beringia," *Proceedings of the National Academy of Science USA* 115 (2018): 4116–4121.

24. Mikkel W. Pedersen, Anthony Ruter, Charles Schweger, et al., "Postglacial Viability and Colonization in North America's Ice-Free Corridor," *Nature* 537, no. 7618 (2016): 45–49, https://doi.org/10.1038/nature19085.

25. References for this section include Charlotte Beck and George T. Jones, "Clovis and Western Stemmed: Population Migration and the Meeting of Two Technologies in the Intermountain West," *American Antiquity* 75, no. 1 (2010): 81–116, https://doi.org/10.7183/0002-7316.75.1.81; Todd J. Braje, Jon M. Erlandson, Torben C. Rick, et al., "Fladmark + 40: What Have We Learned about a Potential Pacific Coast Peopling of the Americas?" *American Antiquity* 85, no. 1 (2020): 1–2, https://doi.org/10.1017/aaq.2019.80.

26. K. R. Fladmark, "Routes: Alternate Migration Corridors for Early Man in North America," *American Antiquity* 44, no. 1 (1979): 55–69, https://doi.org/10.2307/279189; Duggan, Stoneking, Rasmussen et al. (2011); Chris Clarkson, Zenobia Jacobs, Ben Marwick, et al., "Human Occupation of Northern Australia by 65,000 Years Ago," *Nature* 547 (2010): 306–310, https://doi.org/10.1038/nature22968; and Curtis W. Marean, "Early Signs of Human Presence in Australia," *Nature* 547 (2017): 285–286, https://doi.org/10.1038/547285a.

27. Braje, Erlandson, Rick, et al., "Fladmark + 40."

28. For a review of the archaeology and genetics of the peopling of Australia and Oceania, see Ana T. Duggan and Mark Stoneking, "Australia and Oceania," chapter 10 in *Evolution of the Human Genome II*, edited by N. Saitou (Springer Nature, 2021). See also: Chris Clarkson, Zenobia Jacobs, Ben Marwick, et al., "Human Occupation of Northern Australia by 65,000 Years Ago."

29. Jon M. Erlandson, Madonna L. Moss, Matthew Des Lauriers, "Life on the Edge: Early Maritime Cultures of the Pacific Coast of North America," *Quaternary Science Reviews* 27, nos. 23–24 (2008): 2232–2245.

30. Jon M. Erlandson, Todd J. Braje, Kristina M. Gill, et al., "Ecology of the

Kelp Highway: Did Marine Resources Facilitate Human Dispersal from Northeast Asia to the Americas?" *Journal of Island and Coastal Archaeology* 10, no. 3 (2015): 392–411, https://doi.org/10.1080/15564894.2014.100192 3; Braje, Erlandson, Rick, et al., "Fladmark + 40."

31. Clark Larsen, "Biological Distance and Historical Dimensions of Skeletal Variation," chapter 9 in *Bioarchaeology: Interpreting Behavior From the Human Skeleton*, 2nd ed. (Cambridge University Press, 2015), pp. 357–401. I am indebted to Richard Scott and Clark Larsen for comments on this text.

32. G. R. Scott, D. H. O'Rourke, J. A. Raff, et al., "Peopling the Americas: Not 'Out of Japan,'" *PaleoAmerica* 7, no. 4 (2021), 309–332, DOI: 10.1080 /20555563.2021.1940440.

33. Charles Conrad Abbott's diary, February 14, 1877. As quoted on page 68 in Meltzer, *First Peoples in a New World*.

34. L. S. B. Leakey, R. De Ette Simpson, and T. Clements, "Archaeological Excavations in the Calico Mountains, California: Preliminary Report," *Science* 160, no. 3831 (1968): 1022–1023, https://doi.org/10.1126/science.160.3831.1022; V. Haynes, "The Calico Site: Artifacts or Geofacts?" *Science* 181, no. 4097 (1973): 305–310, https://doi.org/10.1126/science.181.4097.305.

35. Steven R. Holen, Thomas A. Deméré, Daniel C. Fisher, et al., "A 130,000-Year-Old Archaeological Site in Southern California, USA," *Nature* 544, no. 7651 (2017): 479–483, https://doi.org/10.1038/nature22065.

36. Katerina Harvati, Carolin Röding, Abel M. Bosman, et al., "Apidima Cave Fossils Provide Earliest Evidence of Homo Sapiens in Eurasia," *Nature* 571, no. 7766 (2019): 500–504, https://doi.org/10.1038/s41586-019-1376-z; Eric Delson, "An Early Modern Human Outside Africa," *Nature* 571 (2019): 487–488, https://doi.org/10.1038/d41586-019-02075-9.

37. Cosimo Posth, Christoph Wissing, Keiko Kitagawa, et al., "Deeply Divergent Archaic Mitochondrial Genome Provides Lower Time Boundary for African Gene Flow into Neanderthals," *Nature Communications* 8, no. 1 (2017): 16046, https://doi.org/10.1038/ncomms16046; Martin Kuhlwilm, Ilan Gronau, Melissa J. Hubisz, et al., "Ancient Gene Flow from Early Modern Humans into Eastern Neanderthals," *Nature* 530, no. 7591 (2016): 429–433, https://doi.org/10.1038/nature16544.

38. Jennifer Raff, "An Extremely Early Migration of Modern Humans Out of Africa," last modified July 11, 2019, https://www.forbes.com/sites /jenniferraff/2019/07/11/an-extremely-early-migration-of-modern -humans-out-of-africa/#6c5c33b9a130.

39. C. Vance Haynes, "Fluted Projectile Points: Their Age and Dispersion: Stratigraphically Controlled Radiocarbon Dating Provides New Evidence on Peopling of the New World," *Science* 145, no. 3639 (1964): 1408–1413.

40. This includes the recently announced find of stone artifacts in sediments dating to 30,000–25,000 years ago from the Chiquihuite Cave site in northern Mexico. Ciprian F. Ardelean, Lorena Becerra-Valdivia, Mikkel Winther Pedersen, et al., "Evidence of Human Occupation in Mexico around the Last Glacial Maximum," *Nature* 584, no. 7819 (2020): 87–92. For an older but thorough review of Pleistocene archaeology in South America, see Tom Dillehay's *The Settlement of the Americas: A New Prehistory* (Basic Books, 2000).

41. Capuchin monkeys of the Serra de Capivara National Park in Brazil have been observed deliberately breaking stones in order to use the tools for food processing, as a source of silica, and to throw in sexual displays. Tomos Proffitt, Lydia V. Luncz, Tiago Falótico, et al., "Wild Monkeys Flake Stone Tools," *Nature* 539, no. 7627 (2016): 85–88, https://doi.org/10.1038/nature20112; Tiago Falótico and Eduardo B. Ottoni, "Stone Throwing as a Sexual Display in Wild Female Bearded Capuchin Monkeys, Sapajus Libidinosus," edited by Michael D. Petraglia, *PLoS ONE* 8, no. 11 (2013): e79535, https://doi.org/10.1371/journal.pone.0079535.

42. Matthew Magnani, Dalyn Grindle, Sarah Loomis, et al., "Evaluating Claims for an Early Peopling of the Americas: Experimental Design and the Cerutti Mastodon Site," *Antiquity* 93, no. 369 (2019): 789–795.

43. Todd J. Braje, Tom D. Dillehay, Jon M. Erlandson, et al., "Were Hominins in California ~130,000 Years Ago?" *PaleoAmerica* 3, no. 3 (2017): 200–202, https://doi.org/10.1080/20555563.2017.1348091.

44. Steven R. Holen, Thomas A. Deméré, Daniel C. Fisher, et al., "Disparate Perspectives on Evidence from the Cerutti Mastodon Site: A Reply to Braje et al.," *PaleoAmerica* 4, no. 1 (2018): 12–15, https://doi.org/10.1080/20555563.2017.1396836; Ruth Gruhn, "Observations Concerning the Cerutti Mastodon Site," *PaleoAmerica* 4, no. 2 (2018): 101–102, https://doi.org/10.1080/20555563.2018.1467192. For an Indigenous archaeologist's perspective on how Cerutti and other early sites can subvert colonialist knowledge production, see Paulette Steeves, *The Indigenous Paleolithic of the Western Hemisphere* (University of Nebraska Press, 2021).

45. A fair summary of the majority archaeological perspectives on aspects of this site is Magnani, Grindle, Loomis, et al., "Evaluating Claims for an Early Peopling of the Americas."

46. See, for example, Graham Hancock's arguments in his book *America Before: The Key to Earth's Lost Civilization* (St. Martin's, 2019).

47. Sriram Sankararaman, Swapan Mallick, Nick Patterson, et al., "The Combined Landscape of Denisovan and Neanderthal Ancestry in Present-Day Humans," *Current Biology* 26, no. 9 (2016): 1241–1247, https://doi.org/10.1016/j.cub.2016.03.037.

48. Pontus Skoglund and David Reich, "A Genomic View of the Peopling of the Americas," *Current Opinion in Genetics and Development* 41 (December 2016): 27–35, https://doi.org/10.1016/j.gde.2016.06.016; Magnani, Grindle, Loomis, et al., "Evaluating Claims for an Early Peopling of the Americas." The Ust'-Ishim genome was published by Qiaomei Fu, Heng Li, Priya Moorjani, et al., "Genome Sequence of a 45,000-Year-Old Modern Human from Western Siberia," *Nature* 514, no. 7523 (2014): 445–449, https://doi.org/10.1038/nature13810.

49. Paulette F. Steeves, "Decolonizing the Past and Present of the Western Hemisphere (the Americas)," *Archaeologies* 11, no. 1 (2015): 42–69, https://doi.org/10.1007/s11759-015-9270-2.

Chapter 3

1. James E. Dixon, "Paleo-Indian: Far Northwest," in *Handbook of North American Indians*, edited by Douglas H. Ubelaker (Smithsonian Institution, 2006), pp. 129–147.

2. Randall Haas, James Watson, Tammy Buonasera, et al., "Female Hunters of the Early Americas," *Science Advances* 6, no. 45 (2020): eabd0310, https://doi.org/10.1126/sciadv.abd0310. As with several subjects in this book, the study of gender in archaeology has an extensive and rich literature. I list here a few references which can serve as a gateway for the interested reader: Philip L. Walker and Della Collins Cook, "Gender and Sex: Viva la Difference," *American Journal of Physical Anthropology* 106 (1998): 255–259; Traci Arden, "Studies of Gender in the Prehispanic Americas," *Journal of Archaeological Research* 16 (2008): 1–35; *Exploring Sex and Gender in Bioarchaeology*, edited by Sabrina C. Agarwal and Julie K. Wesp (University of New Mexico Press, 2017).

3. Willerslev and Meltzer, "Peopling of the Americas As Inferred from Ancient Genomes."

4. See, for example, Katheen Sterling, "Man the Hunter, Woman the Gatherer? The Impact of Gender Studies on Hunter-Gatherer Research (A Retrospective)," chapter 7 in the *Oxford Handbook of the Archaeology and Anthropology of Hunter-Gatherers*, edited by Vicki Cummings, Peter Jordan, and Marek Zvelebil (Oxford University Press, 2014), pp. 151–173; and J. M. Adovasio, Olga Soffer, and Jake Page, *The Invisible Sex: Uncovering the True Roles of Women in Prehistory* (Taylor and Francis, 2007).

5. Charles Holmes, "The Beringian and Transitional Periods in Alaska: Technology of the East Beringian Tradition as Viewed from Swan Point," in *From the Yenisei to the Yukon: Interpreting Lithic Assemblages Variability in*

Late Pleistocene/Early Holocene Beringia, edited by T. Goebel and Ian Buvit (Texas A&M University Press, 2011), pp. 179–191.

6. Including the Dry Creek, Owl Ridge, Walker Road, Moose Creek, Broken Mammoth, Mead, Linda's Point sites. See references in (5) above, and additionally: Kelly E. Graf, Lyndsay M. DiPietro, Kathryn E. Krasinski, et al., "Dry Creek Revisited: New Excavations, Radiocarbon Dates, and Site Formation Inform on the Peopling of Eastern Beringia," *American Antiquity* 80, no. 4 (2015): 671–694.

7. References for include Ben A. Potter, Charles E. Holmes, and David R. Yesner, "Technology and Economy among the Earliest Prehistoric Foragers in Interior Eastern Beringia," in *Paleoamerican Odyssey*, edited by Kelly E. Graf, Caroline V. Ketron, and Michael R. Waters (Texas A&M University Press, 2013), pp. 81–103; Michael R. Bever, "An Overview of Alaskan Late Pleistocene Archaeology: Historical Themes and Current Perspectives," *Journal of World Prehistory* 15, no. 2 (2001): 125–191; Dixon, "Paleo-Indian: Far Northwest"; Frederick H. West, *American Beginnings: The Prehistory and Paleoecology of Beringia* (University of Chicago Press, 1996); Charles E. Holmes, "Tanana River Valley Archaeology circa 14,000 to 9000 B.P.," *Arctic Anthropology* 38, no. 2 (2001): 154–170.

8. See Ben A. Potter, Joshua D. Reuther, Vance T. Holliday, et al., "Early Colonization of Beringia and Northern North America: Chronology, Routes, and Adaptive Strategies," *Quaternary International* 444 (July 2017): 36–55, https://doi.org/10.1016/j.quaint.2017.02.034.

9. References for this section include Yu Hirasawa and Charles E. Holmes, "The Relationship between Microblade Morphology and Production Technology in Alaska from the Perspective of the Swan Point Site," *Quaternary International* 442 (June 2017): 104–117, https://doi.org/10.1016/j.quaint.2016.07.021; Yan Axel Gómez Coutouly, "Pressure Microblade Industries in Pleistocene-Holocene Interior Alaska: Current Data and Discussions," in *The Emergence of Pressure Blade Making*, edited by Pierre M. Desrosiers (Springer, 2012), pp. 347–374; Charles M. Mobley, *The Campus Site: A Prehistoric Camp at Fairbanks, Alaska* (University of Alaska Press, 1991); Kelly E. Graf and Ted Goebel, "Upper Paleolithic Toolstone Procurement and Selection across Beringia," in *Lithic Materials and Paleolithic Societies*, edited by B. Adams, and B. S. Blades (Blackwell Publishing, 2009), pp. 54–77.

10. For discussions, see: Frederick H. West, *American Beginnings: The Prehistory and Paleoecology of Beringia* (University of Chicago Press, 1996); *Paleoamerican Odyssey*, edited by Kelly E. Graf, Caroline V. Ketron, and Michael R. Waters (Texas A&M University Press, 2013), pp. 81–103; Michael R. Bever, "An Overview of Alaskan Late Pleistocene Archaeology: Historical

Themes and Current Perspectives," *Journal of World Prehistory* 15, no. 2 (2001): 125–191.

11. Yan Axel Gómez Coutouly, Angela K. Gore, Charles E. Holmes, et al., "'Knapping, My Child, Is Made of Errors': Apprentice Knappers at Swan Point and Little Panguingue Creek, Two Prehistoric Sites in Central Alaska," *Lithic Technology* 46, no. 1 (2021): 2–26, https://doi.org/10.1080/01977261.2020.1805201; John J. Shea, "Child's Play: Reflections on the Invisibility of Children in the Paleolithic Record," *Evolutionary Anthropology: Issues, News, and Reviews* 15, no. 6 (2006): 212–216, https://doi.org/10.1002/evan.20112.

12. Heather L. Smith and Ted Goebel, "Origins and Spread of Fluted-Point Technology in the Canadian Ice-Free Corridor and Eastern Beringia," *Proceedings of the National Academy of Science USA* 115 (2018): 4116–4121.

13. Frederick Hadleigh West, *The Archaeology of Beringia* (Columbia University Press, 1981).

14. Ben A. Potter, Joshua D. Reuther, Vance T. Holliday, et al., "Early Colonization of Beringia and Northern North America: Chronology, Routes, and Adaptive Strategies," *Quaternary International* 444 (July 2017): 36–55, https://doi.org/10.1016/j.quaint.2017.02.034.

15. For example, Matthew R. Bennett, David Bustos, Jeffrey S. Pigati, et al., "Evidence of Humans in North America during the Last Glacial Maximum," *Science* (2021), in press.

16. For this debate, see Brian Wygal, "The Peopling of Eastern Beringia and Its Archaeological Complexities," *Quaternary International* 466 (2018): 284–298; Frederick West, "The Archaeological Evidence," in American Beginnings: *The Prehistory and Palaeoecology of Beringia*, edited by Frederick West (University of Chicago Press, 1996), pp. 537–560; Charles E. Holmes, "The Beringian Tradition and Transitional Periods in Alaska: Technology of the East Beringian Tradition as Viewed from Swan Point," in *From the Yenisei to the Yukon: Interpreting Lithic Assemblage Variability in Late Pleistocene/Early Holocene Beringia*, edited by Ted Goebel and Ian Buvit (Texas A&M University Press, 2011), pp. 179–191; Ben A. Potter, Charles E. Holmes, and David R. Yesner, "Technology and Economy among the Earliest Prehistoric Foragers in Interior Eastern Beringia," in *Paleoamerican Odyssey*, edited by Kelly Graf, Caroline V. Ketron, and Michael R. Waters (Texas A&M University, 2013), pp. 81–104; Ted Goebel and Ian Buvit, "Introducing the Archaeological Record of Beringia," in *From the Yenisei to the Yukon: Interpreting Lithic Assemblage Variability in Late Pleistocene/Early Holocene Beringia*, edited by Ted Goebel and Ian Buvit (Texas A&M University Press, 2011), pp. 1–30; Michael R. Bever, "An Overview

of Alaskan Pleistocene Archaeology: Historical Themes and Current Perspectives," *Journal of World Prehistory* 15 no. 2 (2001): 125–191; Don E. Dumond, "Technology, Typology, and Subsistence: A Partly Contrarian Look at the Peopling of Beringia," in *From the Yenisei to the Yukon: Interpreting Lithic Assemblage Variability in Late Pleistocene/Early Holocene Beringia*, edited by Ted Goebel and Ian Buvit (Texas A&M University Press, 2011), pp. 345–360.

Chapter 4

1. My family and our club looked on cave divers as a particularly nutty kind of daredevil. I was severely admonished that I would probably die if I ever attempted it. It was unlikely, in Springfield, Missouri, that I ever would attempt it, but I took their warnings to heart.

2. Allen J. Christenson, trans., *Popol Vuh, the Sacred Book of the Quiché Maya People* (University of Oklahoma Press, 2007), https://www.mesoweb.com/publications/Christenson/PopolVuh.pdf.

3. My memories of visiting this cave are a bit shaky after 20 years; I therefore relied upon numerous project reports from the Western Belize Regional Cave Project's 1997 and 1998 reports. In particular, I drew upon these articles: Holley Moyes, and Jaime J. Awe, "Spatial Analysis of Artifacts in the Main Chamber of Actun Tunichil Muknal, Belize: Preliminary Results," in *The Western Belize Cave Project: A Report of the 1997 Field Season*, edited by Jaime J. Awe (University of New Hampshire, 1998); pp. 22–38; Sherry A. Gibbs, "Human Skeletal Remains from Actun Tunichil Muknal and Actun Uayazba Kab," in *The Western Belize Cave Project: A Report of the 1997 Field Season*, edited by Jaime J. Awe (University of New Hampshire, 1998), pp. 71–95.

4. At the time that I visited this cave, I was 20 years old and training at my first archaeological field school—not experienced enough to identify the injuries on the human skeletons or understand their implications. As research for this section, I went back and read the formal archaeological description of the remains. I must confess that I really struggled with writing this section—trying to balance writing sensitively about this topic while providing an accurate description of cultural practices. If I have failed to achieve this balance, I offer my sincere apologies.

5. Jaime J. Awe, Cameron Griffith, and Sherry Gibbs, "Cave Stelae and Megalithic Monuments in Western Belize," in *The Maw of the Earth Monster: Mesoamerican Ritual Cave Use*, edited by James E. Brady and Keith M. Prufer (University of Texas Press, 2005), pp. 223–248.

6. The connection between Abraham's attempted sacrifice of Isaac and the sacrifice of children to Tlaloc has been discussed by Viviana Díaz Balsera, "A Judeo-Christian Tlaloc or a Nahua Yahweh? Domination, Hybridity, and Continuity in the Nahua Evangelization Theater," *Colonial Latin American Review* 10, no. 3 (2001): 209–227, https://doi.org/10.1080/10609160120093787.

7. Michael H. Crawford, *The Origins of Native Americans: Evidence from Anthropological Genetics* (Cambridge University Press, 1998).

8. Cosimo Posth, Nathan Nakatsuka, Iosif Lazaridis, et al., "Reconstructing the Deep Population History of Central and South America," *Cell* 175, no. 5 (2018): 1185–1197, https://doi.org/10.1016/j.cell.2018.10.027.

9. J. Victor Moreno-Mayar, Lasse Vinner, Peter de Barros Damgaard, et al., "Early Human Dispersals within the Americas," *Science* 362, no. 6419 (2018): 1–11, https://doi.org/10.1126/science.aav2621.

Chapter 5

1. Quoted in Charles Petit, "Trying to Study Tribes while Respecting Their Cultures/Hopi Indian Geneticist Can See Both Sides," *SFGate*, February 4, 2012, https://www.sfgate.com/news/article/Trying-to-Study-Tribes-While -Respecting-Their-3012825.php.

2. For privacy reasons, I won't be giving any identifying information about this tribe or project. The results I describe at the end of the chapter are generalized so as to not be identifiable.

3. This has particularly been true for the history of the Americas, where settler colonialism has caused almost incomprehensible harm to the Native peoples throughout the continent, displaced peoples from their homelands, and disconnected many people from their own cultural heritage.

4. Specifically, the part of the mitochondrial genome I was targeting was called hypervariable region I or HVR I for short. There is another hypervariable region in the genome called HVR II, but it's not as informative for distinguishing between lineages. John M. Butler, "Mitochondrial DNA Analysis," in *Advanced Topics in Forensic DNA Typing: Methodology* (San Diego: Academic Press, 2011), pp. 405–456.

5. I was a blundering idiot in his lab—every kind of mistake that is possible to make, I did—but he patiently taught me molecular biology for about six years until I went on to graduate school. As I write this, about three weeks from the official confirmation of achieving tenure, I think about him often and how generous he was to me. Every time I say yes to an undergraduate student wishing to work in my lab, it's because of José.

6. If I could go back and do graduate school over again based on the

knowledge I have now, I would have added bioinformatics training to my toolkit. I hope that maybe someday when I get a sabbatical, I will be able to spend some serious time catching up on all the analytical methods that I never learned how to do myself. If you are an aspiring ancient DNA researcher, I advise you to start learning how to code if possible.

7. The existence of these genomes was one of the clues that led researchers to suspect that mitochondria were once free-living bacteria that somehow developed a symbiotic relationship with other cells.

8. Jennifer A. Raff, Margarita Rzhetskaya, Justin Tackney, et al., "Mitochondrial Diversity of Iñupiat People from the Alaskan North Slope Provides Evidence for the Origins of the Paleo- and Neo-Eskimo Peoples: MtDNA Source of Arctic Migrations," *American Journal of Physical Anthropology* 157, no. 4 (2015): 603–614, https://doi.org/10.1002/ajpa.22750.

9. See https://reich.hms.harvard.edu/cost-effective-enrichment-12-million-snps.

10. For an accessible introduction to these analytical methods, see David Reich, *Who We Are and How We Got Here: Ancient DNA and the New Science of the Human Past* (Oxford University Press, 2018).

Chapter 6

1. References for this section include John F. Hoffecker, *Modern Humans: Their African Origin and Global Dispersal* (Columbia University Press, 2017); K. E. Graf, "Siberian Odyssey," in *Paleoamerican Odyssey*, edited by Kelly E. Graf, Caroline V. Ketron, and Michael R. Waters (Texas A&M University Press, 2013), pp. 65–80; Vladimir Pitulko, Pavel Nikolskiy, Aleksandr Basilyan, and Elena Pavlova, "Human Habitation in Arctic Western Beringia Prior to the LGM," in *Paleoamerican Odyssey*, edited by Kelly E. Graf, Caroline V. Ketron, and Michael R. Waters (Texas A&M University Press, 2013), pp. 13–44; Fu, Li, Moorjani, et al., "Genome Sequence of a 45,000-Year-Old Modern Human from Western Siberia."

2. The SIGMA Type 2 Diabetes Consortium, "Sequence Variants in SLC16A11 Are a Common Risk Factor for Type 2 Diabetes in Mexico," *Nature* 506, no. 7486 (2014): 97–101, https://doi.org/10.1038/nature12828; Reich, *Who We Are and How We Got Here.*

3. References for this section: Maanasa Raghavan, Pontus Skoglund, Kelly E. Graf, et al., "Upper Palaeolithic Siberian Genome Reveals Dual Ancestry of Native Americans," *Nature* 505, no. 7481 (2014): 87–91, https://doi.org/10.1038/nature12736; Martin Sikora, Vladimir Pitulko, Vitor C. Sousa, et al., "The Population History of Northeastern Siberia since the Pleistocene," *Nature* 570 (2019): 182–188; Pontus Skoglund and David

Reich, "A Genomic View of the Peopling of the Americas," *Current Opinion in Genetics and Development* 41 (2016): 27–35.

4. Maanasa Raghavan, Pontus Skoglund, Kelly E. Graf, et al., "Upper Palaeolithic Siberian Genome Reveals Dual Ancestry of Native Americans," *Nature* 505, no. 7481 (2014): 87–91, https://doi.org/10.1038/nature12736.

5. Hoffecker, *Modern Humans*; Vladimir V. Pitulko, Pavel A. Nikolskiy, E. Yu Girya, et al., "The Yana RHS Site: Humans in the Arctic before the Last Glacial Maximum," *Science* 303, no. 5654 (2004): 52–56; Vladimir Pitulko, Pavel Nikolskiy, Aleksandr Basilyan, et al., "Human Habitation in Arctic Western Beringia Prior to the LGM," in *Paleoamerican Odyssey*, edited by Kelly E. Graf, Caroline V. Ketron, and Michael R. Waters (Texas A&M University Press, 2013), pp. 13–44.

6. Kelly E. Graf and Ian Buvit, "Human Dispersal from Siberia to Beringia: Assessing a Beringian Standstill in Light of the Archaeological Evidence," *Current Anthropology* 58, no. S17 (2017): S583–603, https://doi.org/10.1086/693388; K. E. Graf, "Siberian Odyssey," in *Paleoamerican Odyssey*, edited by Kelly Graf, Caroline Ketron, and Michael Waters (Texas A&M University Press, 2013), pp. 65–80.

7. Sikora, Pitulko, Sousa, et al., "The Population History of Northeastern Siberia since the Pleistocene."

8. References for this section include Raghavan, Skoglund, Graf, et al., "Upper Palaeolithic Siberian Genome Reveals Dual Ancestry of Native Americans"; He Yu, Maria A. Spyrou, Marina Karapetian, et al., "Paleolithic to Bronze Age Siberians Reveal Connections with First Americans and across Eurasia," *Cell* 181, no. 6 (2020): 1232–1245.e20, https://doi.org/10.1016/j.cell.2020.04.037.

9. References for this section: Emőke J. E. Szathmáry, "Genetics of Aboriginal North Americans," *Evolutionary Anthropology: Issues, News, and Reviews* 1, no. 6 (2005): 202–20, https://doi.org/10.1002/evan.1360010606; Andrew Kitchen, Michael M. Miyamoto, and Connie J. Mulligan, "A Three-Stage Colonization Model for the Peopling of the Americas," edited by Henry Harpending, *PLoS ONE* 3, no. 2 (2008): e1596, https://doi.org/10.1371/journal.pone.0001596; Erika Tamm, Toomas Kivisild, Maere Reidla, et al., "Beringian Standstill and Spread of Native American Founders," edited by Dee Carter, *PLoS ONE* 2, no. 9 (2007): e829, https://doi.org/10.1371/journal.pone.0000829; Connie J. Mulligan, Andrew Kitchen, and Michael M. Miyamoto, "Updated Three-Stage Model for the Peopling of the Americas," edited by Henry Harpending, *PLoS ONE* 3, no. 9 (2008): e3199, https://doi.org/10.1371/journal.pone.0003199; Thomaz Pinotti, Anders Bergström, Maria Geppert, et al., "Y Chromosome Sequences Reveal

a Short Beringian Standstill, Rapid Expansion, and Early Population Structure of Native American Founders," *Current Biology* 29, no. 1 (2019): 149–157, https://doi.org/10.1016/j.cub.2018.11.029.

10. Sikora, Pitulko, Sousa, et al., "The Population History of Northeastern Siberia since the Pleistocene"; Michael R. Waters, "Late Pleistocene Exploration and Settlement of the Americas by Modern Humans," *Science* 365, no. 6449 (2019): eaat5447, https://doi.org/10.1126/science.aat5447; Jennifer A. Raff, "Genomic Perspectives on the Peopling of the Americas," *SAA Archaeological Record* 19, no. 3 (2019): 12–14; J. F. Hoffecker, S. A. Elias, and D. H. O'Rourke, "Out of Beringia?" *Science* 343, no. 6174 (2014): 979–980, https://doi.org/10.1126/science.1250768.

11. John F. Hoffecker, Scott A. Elias, and Olga Potapova, "Arctic Beringia and Native American Origins," *PaleoAmerica* 6, no. 2 (2020): 158–168, https://doi.org/10.1080/20555563.2020.1725380.

12. Richard S. Vachula, Yongsong Huang, William M. Longo, et al., "Evidence of Ice Age Humans in Eastern Beringia Suggests Early Migration to North America," *Quaternary Science Reviews* 205 (2019): 35–44, https://doi.org/10.1016/j.quascirev.2018.12.003.

13. Heather Pringle, "What Happens When an Archaeologist Challenges Mainstream Scientific Thinking?" *Smithsonian Magazine*, March 8, 2017, https://www.smithsonianmag.com/science-nature/jacques-cinq-mars-bluefish-caves-scientific-progress-180962410; Lauriane Bourgeon, Ariane Burke, and Thomas Higham, "Earliest Human Presence in North America Dated to the Last Glacial Maximum: New Radiocarbon Dates from Bluefish Caves, Canada," edited by John P. Hart, *PLoS ONE* 12, no. 1 (2017): e0169486, https://doi.org/10.1371/journal.pone.0169486.

14. J. Víctor Moreno-Mayar, Ben A. Potter, Lasse Vinner, et al., "Terminal Pleistocene Alaskan Genome Reveals First Founding Population of Native Americans," *Nature* 553, no. 7687 (2018): 203–207, https://doi.org/10.1038/nature25173; Posth, Nakatsuka, Lazaridis, et al., "Reconstructing the Deep Population History of Central and South America"; Sikora, Pitulko, Sousa, et al., "The Population History of Northeastern Siberia since the Pleistocene."

15. Ben A. Potter, Joel D. Irish, Joshua D. Reuther, et al., "New Insights into Eastern Beringian Mortuary Behavior: A Terminal Pleistocene Double Infant Burial at Upward Sun River," *Proceedings of the National Academy of Sciences* 111, no. 48 (2014): 17060–65, https://doi.org/10.1073/pnas.1413131111.

16. See https://news.uaf.edu/oldest-subarctic-north-american-human-remains-found/.

17. Justin C. Tackney, Ben A. Potter, Jennifer Raff, et al., "Two Contempora-

neous Mitogenomes from Terminal Pleistocene Burials in Eastern Beringia," *Proceedings of the National Academy of Sciences* 112, no. 45 (2015): 13833, https://doi.org/10.1073/pnas.1511903112; Moreno-Mayar, Potter, Vinner, et al., "Terminal Pleistocene Alaskan Genome Reveals First Founding Population of Native Americans"; J. Víctor Moreno-Mayar, Lasse Vinner, Peter de Barros Damgaard, et al., "Early Human Dispersals within the Americas," *Science* 362, no. 6419 (2018): eaav2621, https://doi.org/10.1126/science.aav2621; Posth, Nakatsuka, Lazaridis, et al., "Reconstructing the Deep Population History of Central and South America."

18. J. Víctor Moreno-Mayar, Lasse Vinner, Peter de Barros Damgaard, et al., "Early Human Dispersals within the Americas," *Science* 362, no. 6419 (2018): eaav2621, https://doi.org/10.1126/science.aav2621.

19. Angela R. Perri, Tatiana R. Feuerborn, Laurent A. F. Frantz, et al., "Dog Domestication and the Dual Dispersal of People and Dogs into the Americas," *Proceedings of the National Academy of Sciences* 118, no. 6 (2021): e2010083118, https://doi.org/10.1073/pnas.2010083118.

Chapter 7

1. This is an ongoing challenge in archaeology. A site can only be excavated once, and it therefore can only be read about by other archaeologists (unless some portion of the site is preserved for them to study). I believe this has contributed to the extremely diverse array of opinions on which pre-Clovis sites are "good" and which are considered "unreliable." In research for this book I did an informal, completely unscientific polling of six archaeologists whose opinions I regularly rely on to help me interpret genetics results. They didn't have a unanimous opinion about a single site, although most agreed that Page-Ladson was clearly evidence of a pre-Clovis human presence. The sites that most agreed were valid include Swan Point (Alaska), Paisley Caves (Oregon), Shaefer and Hebior (Wisconsin), Monte Verde (Chile), and the Buttermilk Creek sites (Texas).

2. Jessi J. Halligan, Michael R. Waters, Angelina Perrotti, et al., "Pre-Clovis Occupation 14,550 Years Ago at the Page-Ladson Site, Florida, and the Peopling of the Americas," *Science Advances* 2, no. 5 (2016): e1600375, https://doi.org/10.1126/sciadv.1600375.

3. Bastien Llamas, Lars Fehren-Schmitz, Guido Valverde, et al., "Ancient Mitochondrial DNA Provides High-Resolution Time Scale of the Peopling of the Americas," *Science Advances* 2, no. 4 (2016): e1501385, https://doi.org/10.1126/sciadv.1501385.

4. References for this section include Peter Forster, Rosalind Harding,

Antonio Torroni, and Hans-Jurgen Bandelt, "Origin and Evolution of Native American MtDNA Variation: A Reappraisal," *American Journal of Human Genetics* 59, no. 4 (1996): 935–945; Michael D. Brown, Seyed H. Hosseini, Antonio Torroni, et al., "MtDNA Haplogroup X: An Ancient Link between Europe/Western Asia and North America?" *American Journal of Human Genetics* 63, no. 6 (1998): 1852–1861, https://doi .org/10.1086/302155; Antonio Torroni, Theodore G. Schurr, James V. Neel, et al., "Asian Affinities and Continental Radiation of the Four Founding Native American MtDNAs," n.d., 28; Nelson J. R. Fagundes, Ricardo Kanitz, Roberta Eckert, et al., "Mitochondrial Population Genomics Supports a Single Pre-Clovis Origin with a Coastal Route for the Peopling of the Americas," *American Journal of Human Genetics* 82, no. 3 (2008): 583– 592, https://doi.org/10.1016/j.ajhg.2007.11.013; Alessandro Achilli, Ugo A. Perego, Claudio M. Bravi, et al., "The Phylogeny of the Four Pan-American MtDNA Haplogroups: Implications for Evolutionary and Disease Studies," edited by Vincent Macaulay, *PLoS ONE* 3, no. 3 (2008): e1764, https:// doi.org/10.1371/journal.pone.0001764; Ugo A. Perego, Alessandro Achilli, Norman Angerhofer, et al., "Distinctive Paleo-Indian Migration Routes from Beringia Marked by Two Rare MtDNA Haplogroups," *Current Biology* 19, no. 1 (2009): 1–8, https://doi.org/10.1016/j.cub.2008.11.058; Theodore G. Schurr, "The Peopling of the New World: Perspectives from Molecular Anthropology," *Annual Review of Anthropology* 33, no. 1 (2004): 551–583, https://doi.org/10.1146/annurev.anthro.33.070203.143932.

5. Morten Rasmussen, Sarah L. Anzick, Michael R. Waters, et al., "The Genome of a Late Pleistocene Human from a Clovis Burial Site in Western Montana," *Nature* 506, no. 7487 (2014): 225–229, https://doi.org/10.1038 /nature13025.

6. Perri, Feuerborn, Frantz, et al., "Dog Domestication and the Dual Dispersal of People and Dogs into the Americas"; Moreno-Mayar, Potter, Vinner, et al., "Terminal Pleistocene Alaskan Genome Reveals First Founding Population of Native Americans"; J. Víctor Moreno-Mayar, Lasse Vinner, Peter de Barros Damgaard, et al., "Early Human Dispersals within the Americas," *Science* 362, no. 6419 (2018): eaav2621, https://doi.org/10.1126 /science.aav2621; Posth, Nakatsuka, Lazaridis, et al., "Reconstructing the Deep Population History of Central and South America"; Rasmussen, Anzick, Waters, et al., "The Genome of a Late Pleistocene Human from a Clovis Burial Site in Western Montana"; David Reich, Nick Patterson, Desmond Campbell, et al., "Reconstructing Native American Population History," *Nature* 488, no. 7411 (2012): 370–374, https://doi.org/10.1038 /nature11258; C. L. Scheib, Hongjie Li, Tariq Desai, et al., "Ancient

Human Parallel Lineages within North America Contributed to a Coastal Expansion," *Science* 360, no. 6392 (2018): 1024, https://doi.org/10.1126/science.aar6851; Marla Mendes, Isabela Alvim, Victor Borda, et al., "The History behind the Mosaic of the Americas," *Current Opinion in Genetics and Development* 62 (June 2020): 72–77, https://doi.org/10.1016/j.gde.2020.06.007.

7. Pontus Skoglund, Swapan Mallick, Maria Cátira Bortolini, et al., "Genetic Evidence for Two Founding Populations of the Americas," *Nature* 525, no. 7567 (2015): 104–108, https://doi.org/10.1038/nature14895; Reich, *Who We Are and How We Got Here*.

8. Posth, Nakatsuka, Lazaridis, et al., "Reconstructing the Deep Population History of Central and South America"; Marcos Araújo Castro e Silva, Tiago Ferraz, Maria Cátira Bortolini, et al., "Deep Genetic Affinity between Coastal Pacific and Amazonian Natives Evidenced by Australasian Ancestry," *Proceedings of the National Academy of Sciences* 118, no. 14 (2021): e2025739118, https://doi.org/10.1073/pnas.2025739118.

9. Q. Fu, M. Meyer, X. Gao, et al., "DNA Analysis of an Early Modern Human from Tianyuan Cave, China," *Proceedings of the National Academy of Sciences* 110, no. 6 (2013): 2223–2227, https://doi.org/10.1073/pnas.1221359110.

10. References for this section include Ciprian F. Ardelean, Lorena Becerra-Valdivia, Mikkel Winther Pedersen, et al., "Evidence of Human Occupation in Mexico around the Last Glacial Maximum," *Nature* 584, no. 7819 (2020): 87–92; Reich, *Who We Are and How We Got Here*.

Chapter 8

1. Anne M. Jensen, "Nuvuk, Point Barrow, Alaska: The Thule Cemetery and Ipiutak Occupation," PhD diss., Bryn Mawr College, 2009, https://repository.brynmawr.edu/dissertations/26.

2. Scott Elias, *Threats to the Arctic* (Elsevier, 2021).

3. Interested readers can find out more about Anne and her projects at her blog: *Out of Ice: Arctic Archaeology as Seen from Utqiaġvik (Barrow), Alaska*, https://iceandtime.net/author/ajatnuvuk.

   The results from the Nuvuk ancient mitochondrial DNA project can be found in a paper by Justin Tackney, Anne M. Jensen, Carolie Kisielinski, et al., "Molecular Analysis of an Ancient Thule Population at Nuvuk, Point Barrow, Alaska," *American Journal of Physical Anthropology* 168 (2019): 303–317.

4. Justin Tackney and I briefly discuss this history in "A Different Way:

Perspectives on Human Genetic Research from the Arctic," *SAA Archaeological Record* 19, no. 2 (2019): 20–25.

5. The common term used for these peoples has long been "Paleo-E*kimo." However, following the lead of Max Friesen, some Arctic scholars, including our own research group at the University of Kansas and our collaborators (Lauren Norman, Justin Tackney, Dennis O'Rourke, Geoff Hayes, Deborah Bolnick, Austin Reynolds, myself, and a number of other archaeologists), have replaced it with the term *Paleo-Inuit*. As Friesen writes, "the term 'Paleo-E*kimo' maintains the use of the root word 'E*kimo,' which is often now seen as inappropriate given that it is not a self-designation for Inuit and can be considered pejorative in some instances. While no alternative term has gained widespread use in the literature, a logical choice is 'Paleo-Inuit' since this name was advocated by the Inuit Circumpolar Council (Resolution 2010-01), which represents the interests of all Inuit, Iñupiat, Inuvialuit, and Yupik peoples from Siberia to Greenland…Thus, adoption of this term allows archaeologists a rare concrete opportunity to follow the lead of an Inuit organization, rather than relying on 'southern' academic discussions of which terms are appropriate" (Friesen, "Archaeology of the Eastern Arctic," p. 144). I will be using *Paleo-Inuit* and *Inuit* instead of *Paleo-E*kimo* and *Neo-E*kimo* throughout this chapter, except where I am quoting others' work.

6. References for this section include: *The Oxford Handbook of The Prehistoric Arctic*, edited by T. Max Friesen and Owen K. Mason (Oxford University Press, 2016); T. M. Friesen, "Archaeology of the Eastern Arctic," chapter 3 in *Out of the Cold: Archaeology on the Arctic Rim of North America*, edited by Owen K. Mason and T. Max Friesen (SAA Press, 2017); Anne M. Jensen, "Nuvuk, Point Barrow, Alaska: The Thule Cemetery and Ipiutak Occupation"; Ellen Bielawski, "Paleoeskimo Variability: The Early Arctic Small-Tool Tradition in the Central Canadian Arctic," *American Antiquity* (1988): 52–74.

7. Morten Rasmussen, Yingrui Li, Stinus Lindgreen, et al., "Ancient Human Genome Sequence of an Extinct Palaeo-Eskimo," *Nature* 463, no. 7282 (2010): 757–762, https://doi.org/10.1038/nature08835.

8. References for this section include papers in the *Oxford Handbook of the Prehistoric Arctic*, edited by T. Max Friesen and Owen K. Mason (Oxford University Press, 2016) and *Out of the Cold: Archaeology on the Arctic Rim of North America*, edited by Owen K. Mason and T. Max Friesen (SAA Press, 2017).

9. Maanasa Raghavan, Michael DeGiorgio, Anders Albrechtsen, et al., "The Genetic Prehistory of the New World Arctic," *Science* 345, no. 6200 (2014): 1255832, https://doi.org/10.1126/science.1255832.

10. Jennifer Raff, Margarita Rzhetskaya, Justin Tackney, M. Geoffrey Hayes,

"Mitochondrial Diversity of Iñupiat people from the Alaskan North Slope provides evidence for the origins of the Paleo- and Neo-Eskimo Peoples," *American Journal of Physical Anthropology* 157, no. 4 (2015): 603–614.

11. Pavel Flegontov, N. Ezgi Altınışık, Piya Changmai, et al., "Palaeo-Eskimo Genetic Ancestry and the Peopling of Chukotka and North America," *Nature* 570, no. 7760 (2019): 236–240, https://doi.org/10.1038/s41586-019-1251-y; the other study mentioned here was Sikora, Pitulko, Sousa, et al., "The Population History of Northeastern Siberia since the Pleistocene."

12. A cogent discussion of the evidence can be found in Willerslev and Meltzer's "Peopling of the Americas as Inferred from Ancient Genomes," and from Anne C. Stone's "The Lineages of the First Humans to Reach Northeastern Siberia and the Americas," *Nature* 570 (2019): 170–172, https://doi.org/10.1038/d41586-019-01374-5.

13. Don Dumond, "A Reexamination of Eskimo-Aleut Prehistory," *American Anthropologist* 89 (1987): 32–56.

14. Leslea J. Hlusko, Joshua P. Carlson, George Chaplin, et al., "Environmental Selection during the Last Ice Age on the Mother-to-Infant Transmission of Vitamin D and Fatty Acids through Breast Milk," *Proceedings of the National Academy of Sciences* 115, no. 19 (2018): E4426–4432, https://doi.org/10.1073/pnas.1711788115.

15. See William F. Keegan and Corinne L. Hoffman, *The Caribbean Before Columbus* (Oxford University Press, 2017); Scott M. Fitzpatrick, "The Pre-Columbian Caribbean: Colonization, Population Dispersal, and Island Adaptations," *PaleoAmerica* 1, no. 4 (2015): 305–31.

16. References for this section: Scott M. Fitzpatrick, "The Pre-Columbian Caribbean: Colonization, Population Dispersal, and Island Adaptations," *PaleoAmerica* 1, no. 4 (2015): 305–331, https://doi.org/10.1179/2055557115Y.0000000010; Jada Benn Torres, "Genetic Anthropology and Archaeology: Interdisciplinary Approaches to Human History in the Caribbean," *PaleoAmerica* 2, no. 1 (2016): 1–5, https://doi.org/10.1080/20555563.2016.1139859.

17. Kathrin Nägele, Cosimo Posth, Miren Iraeta Orbegozo, et al., "Genomic Insights into the Early Peopling of the Caribbean," *Science* 369, no. 6502 (2020): 456–460; Daniel M. Fernandes, Kendra A. Sirak, Harald Ringbauer, et al., "A Genetic History of the Pre-Contact Caribbean," *Nature* 590, no. 7844 (2021): 103–110, https://doi.org/10.1038/s41586-020-03053-2.

18. Nägele, Posth, Orbegozo, et al., "Genomic Insights into the Early Peopling of the Caribbean"; Daniel M. Fernandes, Kendra A. Sirak, Harald Ringbauer, et al., "A Genetic History of the Pre-Contact Caribbean," *Nature* 590, no. 7844 (2021): 103–110, https://doi.org/10.1038/s41586-020-03053-2.

19. The term *Taíno*, though commonly used colloquially, is controversial. "It's a term first seen in the chronicles and then used by scholars in the 19th and 20th century," Nieves-Colón told me. "Over time it's become the name by which we know these communities, but it's not the term they used to call themselves (that is unknown to us today). The biggest problem with this term is that it's historically been used in a way that homogenizes Indigenous peoples across the region. In the chronicles the Spaniards talk about the Taino of the Greater Antilles as "good, peaceful and noble Indians" and describe them in opposition to the "fierce and cannibal" Caribs of the Lesser Antilles. This false dichotomy probably has a lot to do with the fact that by the 16th century the Crown dictated that only Indigenous peoples who rebelled against colonial rule could be enslaved. Unfortunately, this misnomer continues today in the conceptualization of these peoples reproduced in history books and even by some scholars. L. Antonio Curet has written eloquently about this issue in "The Taíno: Phenomena, Concepts, and Terms," *Ethnohistory* 61, no. 3 (2014): 467–495.

20. Lizzie Wade, "Ancient DNA Reveals Diverse Origins of Caribbean's Earliest Inhabitants," *Science*, June 4, 2020, https://www.sciencemag.org/news/2020/06/ancient-dna-reveals-diverse-origins-caribbean-s-earliest-inhabitants.

21. Nägele, Posth, Orbegozo, et al., "Genomic Insights into the Early Peopling of the Caribbean."

Chapter 9

1. Joseph F. Powell, *The First Americans: Race, Evolution, and the Origin of Native Americans* (Cambridge University Press, 2005).

2. Powell, *The First Americans: Race, Evolution, and the Origin of Native Americans*.

3. James C. Chatters, "The Recovery and First Analysis of an Early Holocene Human Skeleton from Kennewick, Washington," *American Antiquity* 65, no. 2 (2000): 291–316, https://doi.org/10.2307/2694060.

4. David J. Meltzer, *First Peoples in a New World: Colonizing Ice Age America* (University of California Press, 2010); Thomas, *Skull Wars*.

5. Walter A. Neves, Mark Hubbe, and Richard G. Klein, "Cranial Morphology of Early Americans from Lagoa Santa, Brazil: Implications for the Settlement of the New World," *Proceedings of the National Academy of Sciences of the United States of America* 102, no. 51 (2005).

6. J. Víctor Moreno-Mayar, Lasse Vinner, Peter de Barros Damgaard, et

al., "Early Human Dispersals within the Americas," *Science* 362, no. 6419 (2018): eaav2621, https://doi.org/10.1126/science.aav2621.

7. Lizzie Wade, "To Overcome Decades of Mistrust, a Workshop Aims to Train Indigenous Researchers to Be Their Own Genome Experts," *Science* (2018), doi:10.1126/science.aav5286.

8. Douglas W. Owsley and Richard L. Jantz, eds., *Kennewick Man: The Scientific Investigation of an Ancient American Skeleton* (Texas A&M University Press, 2014).

9. Morten Rasmussen, Martin Sikora, Anders Albrechtsen, et al., "The Ancestry and Affiliations of Kennewick Man," *Nature* 523 (2015): 455–458.

10. The repatriation order was an amendment to H.R. 5303, the Water Resources Development Act, Burke Museum, "Statement on the Repatriation of the Ancient One," February 20, 2017, https://www.burkemuseum.org/news/ancient-one-kennewick-man.

11. Jennifer K. Wagner, Chip Colwell, Katrina G. Claw, et al., "Fostering Responsible Research on Ancient DNA," *American Journal of Human Genetics* 107 (2020): 183–195; National Commission for the Protection of Human Subjects of Biomedical and Behavioral Research, *The Belmont Report: Ethical Principles and Guidelines for the Protection of Human Subjects of Research* (Government Printing Office, 1978); Jessica Bardill, Alyssa C. Bader, Nanibaa' A. Garrison, et al., "Advancing the Ethics of Paleogenomics," *Science* 360, no. 6387 (2018): 384–385, https://doi.org/10.1126/science.aaq1131.

12. More accurately, "the child identified on present-day Anzick lands." Rasmussen, Anzick, Waters, et al., "The Genome of a Late Pleistocene Human from a Clovis Burial Site in Western Montana."

13. Douglas J. Kennett, Stephen Plog, Richard J. George, et al., "Archaeogenomic Evidence Reveals Prehistoric Matrilineal Dynasty," *Nature Communications* 8, no. 14115 (2017): 1–9, https://doi.org/10.1038/ncomms14115; Amanda D. Cortez, Deborah A. Bolnick, Jessica Bardill, et al., "An Ethical Crisis in Ancient DNA Research: Insights From the Chaco Canyon Controversy As a Case Study," *Journal of Social Archaeology* (2021). DOI: 10.1177/1469605321991600.

14. Krystal S. Tsosie, Joseph M. Yracheta, Jessica Kolopenuk, et al., "Indigenous Data Sovereignties and Data Sharing in Biological Anthropology," *American Journal of Physical Anthropology* (2020). DOI: 10.1002/ajpa.24184; Krystal S. Tsosie, Joseph M. Yracheta, Jessica A. Kolopenuk, et al., "We Have 'Gifted' Enough: Indigenous Genomic Data Sovereignty in Precision Medicine," *American Journal of Bioethics* 21, no. 4 (2021): 72–75, DOI:

10.1080/15265161.2021.1891347; Krystal S. Tsosie, Keolu Fox, and Joseph M. Yracheta, "Genomics Data: The Broken Promise Is to Indigenous People," *Nature* 591 (2021): 529.

15. Havasupai Tribe, "Welcome to the Official Havasupai Tribe Website," 2017–2020, https://theofficialhavasupaitribe.com/About-Supai/about-supai .html.

16. Paul Rubin, "Indian Givers," *Phoenix New Times*, May 27, 2004, https://www.phoenixnewtimes.com/news/indian-givers-6428347.

17. Nanibaa' A. Garrison and Jessica D. Bardill, "The Ethics of Genetic Ancestry Testing," in *A Companion to Anthropological Genetics*, edited by Dennis O'Rourke (John Wiley & Sons, 2019), pp. 28–29.

18. Kim TallBear, *Native American DNA: Tribal Belonging and the False Promise of Genetic Science* (University of Minnesota Press, 2013).

19. Jessica W. Blanchard, Simon Outram, Gloria Tallbull, et al., "We Don't Need a Swab in Our Mouth to Prove Who We Are," *Current Anthropology* 60, no. 5 (2019).

20. See essay by Joe Yrcheta, "The Warren Debacle Exacerbates a 527 Year Old Problem," https://www.academia.edu/41809911/The_Warren_Debacle _Exacerbates_a_527_year_old_Problem.

21. Darryl Leroux, "'We've Been Here for 2,000 Years': White Settlers, Native American DNA and the Phenomenon of Indigenization," *Social Studies of Science* 48, no. 1 (2018): 80–100.

22. Kim TallBear, "Genomic Articulations of indigeneity," *Social Studies of Science* 43, no. 4 (2013).

23. Hina Walajahi, David R. Wilson, and Sara Chandros Hull, "Constructing Identities: The Implication of DTC Ancestry Testing for Tribal Communities," *Genetics in Medicine* 21 (2019): 1744–1750.

24. Ewan Birney, Michael Inouye, Jennifer Raff, et al., "The Language of Race, Ethnicity, and Ancestry in Human Genetics," arXiv:2106.10041, https://arxiv.org/abs/2106.10041; Daniel J. Lawson, Lucy van Dorp, and Daniel Falush, "A Tutorial on How Not to Over-Interpret STRUCTURE and ADMIXTURE Bar Plots," *Nature Communications* 9, no. 3258 (2018).

25. L. Luca Cavalli-Sforza, "The Human Genome Diversity Project: Past, Present, and Future," *Nature Reviews Genetics* 6, no. 4 (2005): 333–340.

26. The 1000 Genomes Project Consortium, "A Global Reference for Human Genetic Variation," *Nature* 526 (2015): 68–74.

27. Discussed in Nanibaa' A. Garrison, Māui Hudson, Leah L. Ballantyne, et al., "Genomic Research through an Indigenous Lens: Understanding the Expectations," *Annual Review of Genomics and Human Genetics* 20 (2019): 495–517, https://doi.org/10.1146/annurev-genom-083118-015434.

28. Jenny Reardon, "Decoding Race and Human Difference in a Genomic Age," *Differences: A Journal of Feminist Cultural Studies* 15, no. 3 (2004): 38–65, http://muse.jhu.edu/article/174491.

29. M. Dodson and R. Williamson, "Indigenous Peoples and the Morality of the Human Genome Diversity Project," *Journal of Medical Ethics* 25, no. 2 (1999): 204–208, https://doi.org/10.1136/jme.25.2.204.

30. K. G. Claw, M. Z. Anderson, R. L. Begay, et al., "A Framework for Enhancing Ethical Genomic Research with Indigenous Communities," *Nature Communications* 9, no. 2957 (2018), https://doi.org/10.1038/s41467-018-05188-3.

31. At the time of this writing, this includes 45 from Mexico, two from Canada (Athabaskan), one from the US (Aleutian), and two from Greenland (Inuit). At the time of this writing, ancient genomes include Upward Sun River, Saqqaq, Anzick, three from Nevada (Spirit and Lovelock caves), and one from California (San Clemente Island).

32. Elizabeth Weiss and James W. Springer, *Repatriation and Erasing the Past* (University of Florida Press, 2020).

33. Māui Hudson, Nanibaa' A. Garrison, Rogina Sterling, et al., "Rights, Interests, and Expectations: Indigenous Perspectives on Unrestricted Access to Genomic Data," *Nature Reviews Genetics* 21, no. 4863 (2020): 377–384; Keolu Fox, "The Illusion of Inclusion—the 'All of Us' Research Program and Indigenous Peoples' DNA," *New England Journal of Medicine* 383 (2020): 411–413, DOI: 10.1056/NEJMp1915987.

34. Sara Reardon, "Navajo Nation Reconsiders Ban on Genetic Research," *Nature* 550 (2017): 165–166, https://www.nature.com/news/navajo-nation-reconsiders-ban-on-genetic-research-1.22780; Jennifer Q. Chadwick, Kenneth C. Copeland, Dannielle E. Branam, et al., "Genomic Research and American Indian Tribal Communities in Oklahoma: Learning from Past Research Misconduct and Building Future Trusting Partnerships," *American Journal of Epidemiology* 188, no. 7 (2019): 1206–1212, https://doi.org/10.1093/aje/kwz062.

35. Jessica Bardill, Alyssa C. Bader, Nanibaa' A. Garrison, et al., Summer Internship for Indigenous Peoples in Genomics (SING) Consortium, "Advancing the Ethics of Paleogenomics," *Science* 360, no. 6387 (2018): 384–385; Nanibaa' A. Garrison, Māui Hudson, Leah L. Ballantyne, et al., "Genomic Research through an Indigenous Lens: Understanding the Expectations," *Annual Review of Genomics and Human Genetics* 20 (2019): 495–517, https://doi.org/10.1146/annurev-genom-083118-015434; Jennifer K. Wagner, Chip Colwell, Katrina G. Claw, et al., "Fostering Responsible Research on Ancient DNA," *American Journal of Human Genetics* 107 (2020): 183–195.

36. Ana T. Duggan, Alison J.T. Harris, Stephanie Marciniak, et al., "Genetic Discontinuity between the Maritime Archaic and Beothuk Populations in Newfoundland, Canada," *Current Biology* 27, no. 20 (2017): 3149–3156.

Epilogue

1. Keolu Fox and John Hawks, "Use Ancient Remains More Wisely," *Nature* 572 (2019): 581–583, doi: https://doi.org/10.1038/d41586-019-02516-5.

Index

Page numbers in *italics* refer to figures/images.

Image Credits

Interior Images

Photo Insert

About the Author

Dr. Jennifer Raff is an anthropological geneticist and science writer. She studied molecular, cellular, and developmental biology and biological anthropology at Indiana University, earning a dual-major PhD in the Biology and Anthropology departments before doing postdoctoral research at the University of Utah, Northwestern University, and the University of Texas. She is currently an associate professor of Anthropology and affiliate faculty with the Indigenous Studies Program, working with tribes and communities across North America to use ancient and contemporary genomes as tools for investigating historical questions. Her research focuses on the initial peopling of the Americas as well as more recent histories in the North American Arctic and mid-continent. She has written for the public on genetics, history, race, and science literacy at her blog, *Violent Metaphors*, the *New York Times*, *The Guardian*, *Scientific American*, *Huffington Post*, the Evolution Institute, and *Forbes*. She lives in Lawrence, Kansas.